Principles of Natural Lighting

ELSEVIER ARCHITECTURAL SCIENCE SERIES

Editor

HENRY J. COWAN

Professor of Architectural Science
University of Sydney

Previously published

An Historical Outline of Architectural Science
by H. J. COWAN

Thermal Performance of Buildings
by J. F. VAN STRAATEN

Computers in Architectural Design
by D. CAMPION

Fundamental Foundations
by W. FISHER CASSIE,

Models in Architecture
by H. J. COWAN, G. D. DING, J. S. GERO and R. W. MUNCEY

In press

Electrical Services in Buildings
by P. JAY and J. HEMSLEY

Principles
of
Natural Lighting

by

J. A. LYNES
B.SC.(Eng.), C.Eng., M.I.E.E., F.I.E.S.

Lecturer of Architecture
University of Manchester

ELSEVIER PUBLISHING COMPANY LTD
AMSTERDAM – LONDON – NEW YORK
1968

ELSEVIER PUBLISHING COMPANY LTD
BARKING, ESSEX, ENGLAND

ELSEVIER PUBLISHING COMPANY
335 JAN VAN GALENSTRAAT, P.O. BOX 211, AMSTERDAM
THE NETHERLANDS

AMERICAN ELSEVIER PUBLISHING COMPANY INC.
52 VANDERBILT AVENUE, NEW YORK, N.Y. 10017

444-20030-4

LIBRARY OF CONGRESS CATALOG CARD NUMBER 68-24661

WITH 106 ILLUSTRATIONS AND 3 TABLES

Printed in Great Britain by Galliard Limited, Great Yarmouth, England

Preface

Daylight is today one of the growth-points in lighting research. Though the needs of students and of practising architects are well met by Walsh's *The Science of Daylight* and by Hopkinson's *Daylighting*, there is still little guidance for those about to embark upon research in natural lighting. It is to help such people that this small volume contains a longish chapter on instrumentation. For their sake too this book refrains from cataloguing every one of the multitude of prediction techniques available; instead two have been dealt with in some depth—the lumen method for rooflights and gnomonic projection for side windows. These should enable the reader to solve the vast majority of daylight problems. Once he has grasped the underlying principles he can decide for himself which of the many other techniques best meets his needs. The great merit of the gnomonic projection is that it enables the designer to plan simultaneously for daylight and for sunlight. Sunlight is dealt with in a companion volume in this series—Professor Markus' *The Sun and Building Design*. These two topics are virtually inseparable; obviously the two books should be read together.

Those tables and charts which are likely to be used most frequently have been grouped together in an appendix for easy reference. A handful of numerical questions and answers are included to encourage readers to "have a go". Answers are deliberately quoted to an unrealistic degree of accuracy; this is to help readers to check their own results, not to suggest that great precision is warranted in daylight calculations.

It is impossible to mention by name all who have helped me to write this book. I share with all my generation of lighting engineers an enormous debt to Mr. J. M. Waldram and to Professor R. G. Hopkinson.

My treatment of the illumination vector is largely due to the late Professor A. A. Gershun of Leningrad, whose work is still barely recognised in the English-speaking world. Mr. W. H. Stephenson tore an early draft of Chapter 6 to shreds: if the present version does not meet his exacting standards the fault is mine, not his.

On a more personal level I have to thank my old chief, Professor T. A. Markus, for triggering off my interest in natural lighting, and Peter Jay for making me doubt everything.

I am also indebted to my former colleagues of the Pilkington Daylight Advisory Service for their patience, which I continue to try repeatedly, and especially to Kit Cuttle and Ken Wilson, for preparing most of the illustrations, and to Kath Bradbury, for deciphering and typing the manuscript.

Finally, especial thanks are due to the Directors of Pilkington Brothers Limited who have provided every possible facility for the preparation of this book.

The following have kindly permitted the reproduction of illustrations or tables:

The National Physical Laboratory (Figures 5.1 (a) and 5.1 (b)).
The Illuminating Engineering Society, New York (Figure 4.3).
The Illuminating Engineering Society, London (Tables A.8.III and A.10).

Contents

Chapter 1

Lightness and Luminosity

1.1. Lightness Constancy

In total darkness we cannot see. A source of light, such as the flame of a candle, becomes visible when rays of light travel straight from the source to our eyes. Most objects are not self-luminous; we see them only by reflected light. The amount and the colour of the light which reaches our eyes from such an object depend both on the sources which illuminate it—they may have different spectra, different intensities, etc.—and on the texture and reflective properties of the object itself. A dark object under strong illumination may reflect the same amount of light as does a white object under weak illumination; if so the light reaching our eyes will be the same in each case, and if our eyes behaved like a television camera we would presumably be unable to distinguish between the two objects.

However, it is a matter of common experience that we generally recognise a sheet of paper as white, or a lump of coal as black, in sunshine or by the light of the moon, despite the fact that the coal in sunlight reflects much more light than does the paper by moonlight. We have learned to distinguish between the effects of the lighting and the characteristics of the surfaces on which the illumination falls; we almost never mistake black objects for white objects, or even light-grey objects for dark-grey objects. Our ability to tell them apart is known as *lightness constancy*. Clearly this effect plays an important part in enabling us to recognise the things around us. It should, however, be regarded as a tendency rather than as an unbreakable law. We cannot invariably judge the lightness of an object accurately and indeed the study of natural lighting is concerned as much with the limitations of lightness constancy as with the phenomenon itself.

The tendency towards lightness constancy is strengthened when the thing we are looking at has a clearly perceived shape, texture and orientation, and when it is seen against a familiar background or

1

among familiar objects with distinctive surfaces. Conversely lightness constancy may break down completely if the object is viewed through a narrow black tube which excludes the edges of the object. In this case we may see the brightness of the aperture rise or fall, but cannot tell if this betokens a change in the illumination or a change in the greyness of the surface itself. Lightness constancy can also be impaired in less unusual circumstances. A floodlit building at night can acquire a self-luminous quality which defies us to judge whether it is white or grey. A gloomy room is another example where lightness constancy is diminished and we cannot judge the lightness of room surfaces with any confidence. Sometimes a lighting designer will deliberately exploit the breakdown of constancy to obtain a dramatic effect. More usually he will be aiming for accurate vision and the avoidance of gloom, and the preservation of constancy will be an important criterion in good lighting [1.1].

The phenomenon of lightness constancy, and the consequent artificiality of any one-to-one relationship between the radiation reaching the eye and the impression of luminosity which it produces, constitutes a paradox at the heart of illuminating engineering. Although a physicist can easily measure or calculate the radiation which an object reflects, he cannot judge how bright it will appear to a human observer. The perceived luminosity will depend on the object's relationship to its surroundings. Indeed if no surroundings can be seen—if the view presented to the eye is totally uniform and undifferentiated—no light is perceived at all. This condition prevails when halved table-tennis balls are fitted over one's eyes; after less than a minute vision fails and is restored only by a reflex jerk of the eyeball breaking the uniform field.

The ultimate theoretical aim of photometry—to devise an instrument which will measure lighting in terms of its visual effect—is thus unlikely ever to be achieved in situations which have much interest for the lighting designer. Indeed there is no *a priori* reason to suppose that it is possible to measure physically any of the magnitudes which we perceive. Similar difficulties occur in the study of taste, warmth, noise, etc. Generally the best that can be achieved is to construct tentative scales of measurement, and then find experimentally the range of circumstances, if any, within which they are valid, i.e., the range within which such arithmetical operations as addition, multiplication or integration to the numbers derived would yield predictions consistent with what we actually perceive.

Before photoelectric cells became available all photometric measurements relied ultimately upon visual judgements. As a consequence of lightness constancy even a skilled lighting engineer cannot estimate the illumination in a well-lit room with any confidence; errors of 50 per cent are quite common. The one photometric judgement that human vision can achieve within about 1 per cent is to assess whether two patches of light of the same colour appear equally bright. This order of accuracy can be attained only by eliminating the effects of lightness constancy, i.e., by deliberately concealing all clues as to the sources of the light or the nature of the illuminated surfaces. Until about forty years ago all photometric instruments had to be designed to provide these conditions of viewing. To compare the intensities of two sources the apparatus was arranged so that each source produced a separate patch of light, and the two patches were viewed as close together as possible, against a common background. The brightness of one patch was then adjusted, e.g., by altering its distance from the light source, until it looked as bright as the other patch. By noting the relative distances of the two sources one could then compare their intensities; the procedure is explained in Chapter 2.

The design and use of photometric instruments based on these principles are discussed in Chapter 6. It will be noted that in visual photometry the observer is assumed to be unable to tell whether object A looks, say, twice as bright as object B. The apparatus is always arranged so that patches of light can be adjusted until they appear equally bright [1.2].

1.2. The Relative Luminous Efficiency Function

If the two light sources are of different colours, say red and green, it is even harder to compare their intensities. The technique described above yields errors of about ten per cent. There are two ways of solving the problem: the step-by-step (or cascade) method, where the colour difference is bridged by matching with each other a succession of sources of intermediate colour, and the flicker method in which the two patches of light are viewed in rapid succession and are assumed to be equally bright when the impression of flicker is minimised.

When patches of monochromatic light of different wavelength are compared by these two methods it is found that equally bright patches do not necessarily contain equal amounts of radiation. A yellow patch, for example, is always found to radiate less energy than an equally bright red patch. In 1924 the Commission Internationale de l'Éclairage* agreed on a curve of *relative luminous efficiency*, $V(\lambda)$ (Fig. 1.1), embodying the results of several investiga-

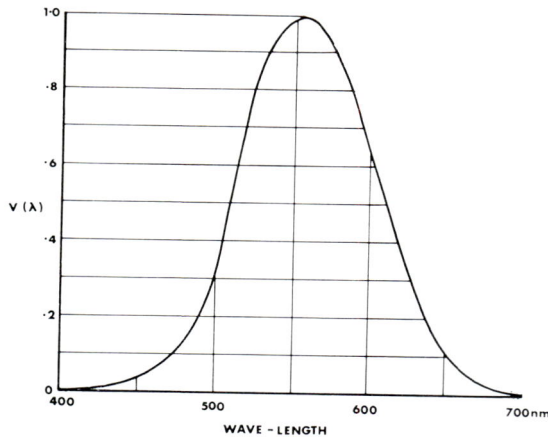

Fig.1.1.   The C.I.E. relative luminous efficiency curve. (1 nm $= 10^{-9}$ metre).

tions of this type [1.3]. The maximum value was set arbitrarily at 1·0, and other values of $V(\lambda)$ were adjusted so that if E_1 watts of radiation at wavelength λ_1 matched E_2 watts of radiation at wavelength λ_2

$$V(\lambda_1) \times E_1 = V(\lambda_2) \times E_2.$$

---

* The Commission Internationale de l'Éclairage, better known by its initials as the C.I.E., is an international clearing-house for decisions on photometry, colorimetry and lighting practice. The address of its Central Bureau is: 25 Rue de la Pépinière, 75–Paris 8, France.

Figure 1.1 thus defines the photometric characteristics of a fictitious *C.I.E. Standard Observer*, whose performance provides an agreed basis for comparing different sources of light. Results would be affected in practice by:

1. The size of the patches of light; the C.I.E. data were obtained for patches subtending 2 to 3 degrees at the eye.

2. The part of the retina illuminated; the C.I.E. data apply to the centre of the visual field, close to the fovea.

3. Adaptation effects. When the observer is adapted to a lower brightness he becomes relatively more sensitive to blue light and less sensitive to red light. This effect, known as the *Purkinje effect*, applies only to luminances below 3 foot-lamberts and can generally be ignored in daylighting studies. However, it probably does affect the C.I.E. relative visibility curve which was based upon experiments carried out at lower adaptation levels.

4. The measurement technique. Although the C.I.E. data were based upon both flicker and step-by-step measurements there may well be systematic differences between results obtained by the two techniques: in the first case the two patches are thrown alternately on to the same portion of the retina; in the second case two separate areas of the retina are involved, in different states of chromatic adaptation [1.4].

5. Individual variations. Significant differences have been found between the relative luminous efficiency curves of observers having normal colour vision. Older people are generally less sensitive to blue light. Furthermore the shape of any one individual's relative luminous efficiency curve seems to vary from season to season. This may well be due to seasonal changes in vitamin intake, for it is known that vitamin A is required for the synthesis of certain photosensitive pigments in the retina.

This recital of the uncertainties surrounding the relative visibility curve serves to emphasise the arbitrary basis of photometry and, for that matter, of almost all psycho-physical measurement. This is well understood among photometrists, but sometimes leads the unwary to "discover" discrepancies between light-meter readings and visual perception. In fact it would have been impossible to construct a scale for measuring lights of different colours without first postulating a fixed relative efficiency curve, and then finding experimentally how far the results predicted by the curve are borne out in practice. In practice the C.I.E. $V(\lambda)$ curve generally gives acceptable results when lightness constancy effects are small and is therefore universally accepted among lighting engineers. When, as normally happens, lightness constancy is important we cannot expect close agreement between photometry and visual perception.

Chapter 2

The Small Light Source

2.1. Intensity

Radiation from a light source can be measured by means of a thermopile. The radiant power (or radiant flux) emitted by the source would then be expressed in watts. Since most sources of visible light also generate invisible radiation—infrared and ultraviolet—which would also be recorded by the thermopile, this instrument cannot readily be used for measuring how much light is emitted by different sources of radiation.

An ideal *photometer* would respond to monochromatic radiant power $P(\lambda)$ of any wavelength λ in proportion to the relative luminous efficiency $V(\lambda)$ for that wavelength, and would also integrate this weighted radiant energy over the whole spectrum:

$$\text{Light flux} = K_{\mathrm{m}} \ P(\lambda) \ V(\lambda) \ \mathrm{d}\lambda. \tag{2.1}$$

This idealised photometer response can be used to define the unit of light flux—the *lumen*—once the value of K_{m} is known. K_{m} is equal to the number of lumens in one watt of monochromatic light at the wavelength of the peak of the $V(\lambda)$ curve—680 lumens per watt. By putting this value in eqn. (2.1) we obtain the following definition

$$\text{Light flux (in lumens)} = 680 \int P(\lambda) \ V(\lambda) \ \mathrm{d}\lambda. \tag{2.2}$$

Equation (2.2) provides a workable definition for lighting calculations, but for the purposes of a standardising laboratory the lumen has been defined in more concrete terms as one-sixtieth of the peak value of light flux per steradian emitted by one square centimetre of a Planckian radiator at the temperature of solidification of molten platinum. (A steradian is simply a measure of the solid angle subtended at a point by a surface. If we imagine the point at the centre of a sphere, the solid angle in steradians subtended by any surface is equal to the area it subtends on the surface of the sphere, divided by the square of the sphere radius. An unobstructed hemi-

7

sphere of sky would subtend 2π steradians; a small object of subtended area A at a distance d would subtend A/d^2 steradians.)

When light flux strikes a surface the *illumination* is measured in
terms of the flux incident per unit of surface area, e.g., in lumens per
square foot. Various units of illumination are listed in the Appendix
(Table A.2.I).

Since light is a form of radiant energy, the principle of conservation of energy requires that illumination shall be additive, i.e., the
illumination received, by each element on a surface, from two or more
light sources must be equal to the sum of the illuminations received
from each source independently even when the sources differ in
colour. This *Law of Additivity* (sometimes known as Abney's Law) is
one of the basic laws of illumination. Two other fundamental laws
may be stated as follows:

If an illumination E_1 matches an illumination E_3, which may be
of a different colour, and if an illumination E_2 also matches E_3,

then $E_1 = E_2$ (the *Law of Transitivity*)

and $mE_1 + nE_2 = E_3$ (the *Law of Distributivity*)

where $m + n = 1$.

The other two basic laws of illumination, the Inverse Square
Law and the Cosine Law, both follow from the fact that light
travels in straight lines.

In Fig. 2.1, O is a small light source. The surfaces at B and C
both subtend the same solid angle at O, so they would both intercept the same light flux (lumens) from O, but since their areas are

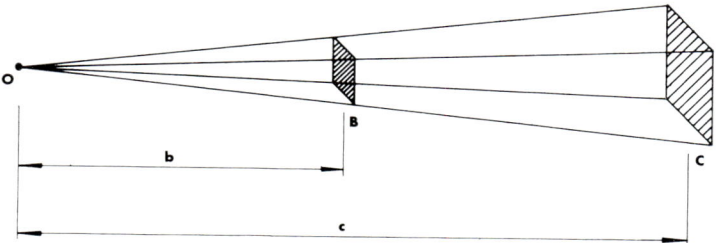

Fig. 2.1.   The inverse square law.

different they would not receive the same illumination (lumens per square foot). Since their areas are proportional to the square of their distance from O, the illumination, E, on each surface must be inversely proportional to the square of its distance from the light source. This *Inverse Square Law* is expressed in the form

$$c^2 \times E_C = b^2 \times E_B = I.$$

The constant I is known as the *intensity* of the source O in the direction of C and B. It is expressed in *candelas*. Alternatively we may write:

$$\text{Illumination } E = I/d^2 \tag{2.3}$$

where d = distance from source to surface.

In Fig. 2.1 the rays of light are perpendicular to the receiving surface. If the surface is turned through an angle θ, as in Fig. 2.2,

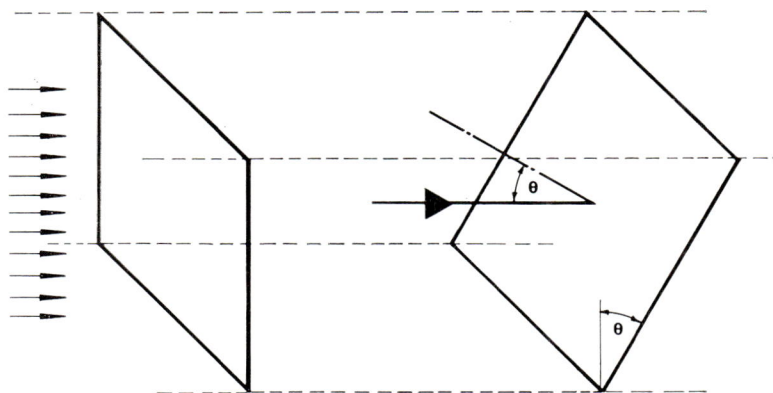

Fig. 2.2. The cosine law of illumination.

so that the rays fall obliquely, the same flux is spread over a larger area. This area will be inversely proportional to the cosine of the incident angle θ, so the illumination (lumens per unit area) will be proportional to cos θ. This *Cosine Law* may be combined with the inverse square law (eqn. (2.3)):

$$E = \frac{I \cos \theta}{d^2}. \tag{2.4}$$

The inverse square law, and the concept of intensity which depends upon it, apply strictly only for a point source of light. This was an acceptable postulate in the eighteenth century, when the foundations of photometry were evolved by scientists whose principal light source, for experimental purposes, was a candle, and it is hardly surprising that some of their conceptual tools are poorly adapted to the study of natural lighting in the twentieth century. The inverse square law cannot be applied with perfect accuracy to a finite light source, such as a window, unless the dimensions of the source are negligible when compared with its distance. However, the inverse square law is a useful approximation at much smaller distances, and can be applied to a diffuse light source (i.e., not comprising lenses, prisms, specular reflectors or louvers) provided that the distance, which should be measured from the centroid of the source, exceeds the maximum dimension, usually the diagonal, of the source [2.1].

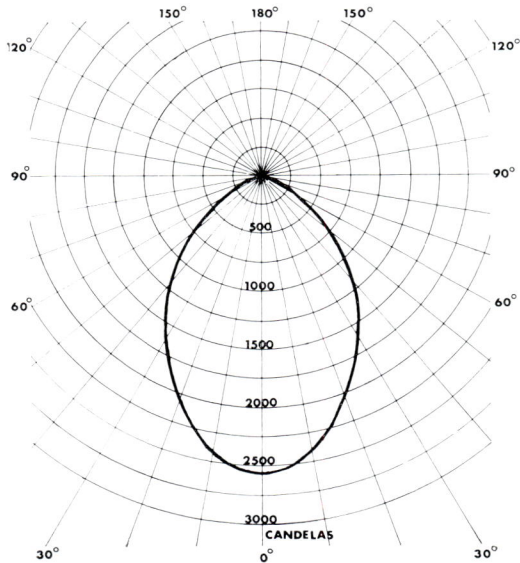

Fig. 2.3.   Polar curve of intensity from 3-foot diameter roof dome on 9-inch white skirt, beneath an overcast sky illumination of 1000 lumens per square foot.

The performance of a source of light can be completely described by plotting the directional distribution of its intensity, provided that its size and position are such that the inverse square law can be applied without serious error.

Figure 2.3 shows a typical *polar curve* for a glass dome rooflight. This is a graph of the intensity distribution plotted in polar co-ordinates as a function of the angle θ measured from the vertical axis of symmetry.

2.2. The Illumination Vector

Figure 2.4 (a) shows an element of a plane surface illuminated nor-mally by a point source O having an intensity I_O in the direction of the element. If the element is now turned through an angle θ its illumina-tion will decrease in accordance with the Cosine Law (eqn. (2.4)). In Fig. 2.5 (a) the illumination E is plotted as a function of the angle θ through which the element has been turned. When θ lies between 90 degrees and 270 degrees the illumination is seen to be less than zero, for eqn. (2.4) implies the convention that light striking the back surface of the element is measured as negative illumination.

A second light source, for example that shown at Q in Fig. 2.4 (a), will produce a second sine wave, as in Fig. 2.5 (b), of the same

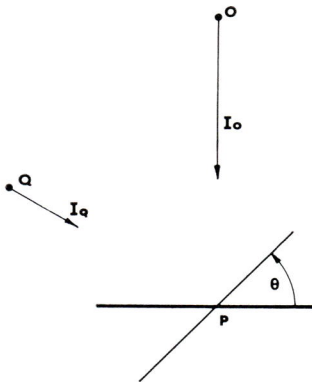

Fig. 2.4 (a).   Rotation of plane element.

period but, in general, displaced laterally and with a different peak value.

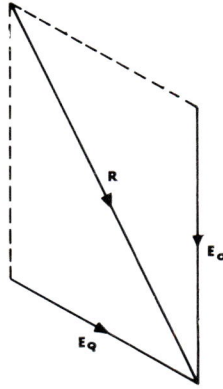

Fig. 2.4 (b).   Parallelogram of vectors.

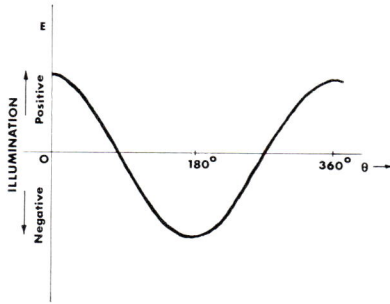

Fig. 2.5 (a).   Variation of illumination as element is turned.

It may be shown that the algebraic sum of two sine waves of a given frequency is a third sine wave having the same frequency:

Let any two sine waves of frequency ω be represented by the equations

$$x_1 = a \sin (\omega t + \alpha) \tag{2.5}$$

$$x_2 = b \sin (\omega t + \beta). \tag{2.6}$$

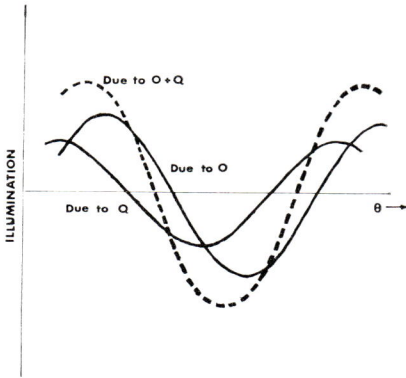

Fig. 2.5 (b). Algebraic summation of sine waves.

Their sum $(x_1 + x_2)$ is obtained by superposition

$$x_1 + x_2 = a \sin (\omega t + \alpha) + b \sin (\omega t + \beta)$$
$$x_1 + x_2 = (a \cos \alpha + b \cos \beta) \sin \omega t$$
$$+ (a \sin \alpha + b \sin \beta) \cos \omega t.$$

Put $\quad R = \{(a \cos \alpha + b \cos \beta)^2 + (a \sin \alpha + b \sin \beta)^2\}^{\frac{1}{2}}$ (2.7)

and let $\qquad A = \tan^{-1} \dfrac{a \sin \alpha + b \sin \beta}{a \cos \alpha + b \cos \beta}$ (2.8)

$$x_1 + x_2 = R(\cos A \sin \omega t + \sin A \cos \omega t)$$
$$x_1 + x_2 = R \sin (\omega t + A). \tag{2.9}$$

Equation (2.9) shows that the sum $(x_1 + x_2)$ of the sine waves represented by eqns. (2.5) and (2.6) is a third sine wave having the same frequency ω. Equations (2.7) and (2.8) show respectively the amplitude R of the resultant and its angular displacement A. It will be seen that this resultant is the vectorial sum of eqns. (2.5) and (2.6).

The algebraic sum of the illumination distribution curves for O and Q is shown dotted in Fig. 2.5 (b). Since this is the vector resultant of the illumination due to the two sources, its magnitude and direction can be obtained graphically (Fig. 2.4 (b)) by drawing vectors at the point P representing, in magnitude and direction, the normal illumination from O and Q, calculated from eqn. (2.3). The arrows on the vectors point away from the light source in each case. The magnitude and direction of the resultant R are obtained by completing the parallelogram of vectors; this shows the amplitude and phase of the resultant sine wave.

If we now add the illumination from a third point source, not necessarily in the same plane, this can be combined, by vectorial addition, with the resultant of the first two vectors to produce a further resultant. We can imagine this combined with the resultant of other point sources, an infinite number if need be. Since a large source is merely an infinite number of point sources it must be possible, in principle, to obtain the resultant vector at any point in space for a surface source, an assembly of surface sources, and for the complex pattern of window, lamps and reflective surfaces visible from any given point in any room. This resultant vector is known as the *illumination vector*, $\overrightarrow{E}$.

If the normal to an element at P (Fig. 2.6) makes an angle θ with the direction of the illumination vector $\overrightarrow{E}$, and the illuminations on the two sides of the element are E_1 and E_2, then

$$E_1 - E_2 = \overrightarrow{E} \cos \theta. \qquad (2.10)$$

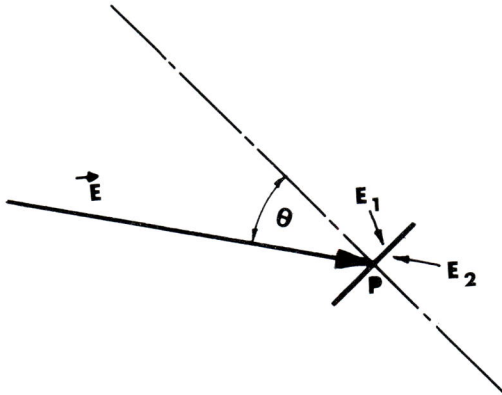

Fig. 2.6.   The illumination vector.

The cosine law of illumination for parallel rays of incident light can be regarded as a particular case of the more general law expressed by eqn. (2.10). If one side of the element is in the dark, i.e., if $E_2 = 0$, $E_1 = \overrightarrow{E} \cos \theta$, so the cosine law applies as much to light from extended sources as to illumination from point sources, so long as the whole source is on the same side of the element.

If the element is turned so that θ equals zero, i.e., the plane of the element is normal to the direction of the illumination vector,

$$E_1 - E_2 = \overrightarrow{E}.$$

The illumination vector $\overrightarrow{E}$ at a point can therefore be defined as the maximum possible difference $(E_1 - E_2)_{max}$ between the illumination E_1 at the front and the illumination E_2 at the back of a plane element at that point [2.2].

It is important to note that although the illumination vector is measured in illumination units—lumens per square foot, etc.—it is equal to an illumination *difference*. Two illumination vectors can be added to produce a smaller vector. This does not contravene the principle of conservation of energy. The true illuminations are always added arithmetically; it is only the illumination *vector* which can be added vectorially.

2.3. The Calculation of Illumination

In Fig. 2.7 the rooflight at O has an intensity $I(\phi)$ in the direction of the point P. Neglecting any light that may be reflected by the walls

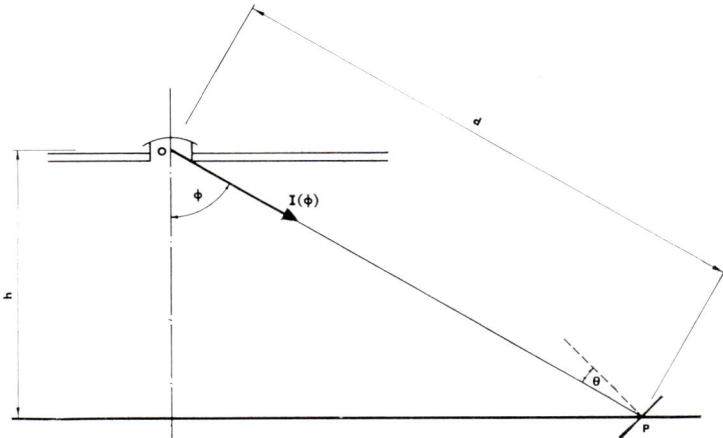

Fig. 2.7. Illumination from a small source.

or floor of the room we may find the illumination vector $\overrightarrow{E}$ at P by applying the inverse square law.

$$\overrightarrow{E} = \frac{I(\phi)}{d^2} \qquad (2.11)$$

The illumination E on any plane at P whose normal is inclined at an angle θ to the line OP will be

$$E = \overrightarrow{E} \cos \theta.$$

For a horizontal plane at P, $\theta = \phi$, so the horizontal illumination E_H is

$$E_H = \overrightarrow{E} \cos \phi = \frac{I(\phi) \cos^3 \phi}{h^2}. \qquad (2.12)$$

Equation (2.12) is known as the *Cosine-cubed Rule*. With the aid of a polar curve showing $I(\phi)$ and a table of $\cos^3 \phi$ (*see* Table A.2.II in Appendix) we can use this expression to estimate the horizontal illumination at any given distance from a light source the height of which is known.

If we are concerned with the lighting of a solid object at P the illumination on one particular plane is only of limited interest. Instead we can calculate the *mean spherical illumination* or *scalar illumination E_s*. The scalar illumination at P is defined as the average illumination over the surface of an infinitesimal sphere situated at P [2.2].

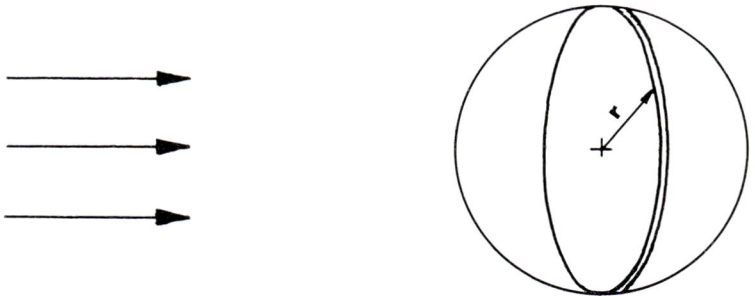

Fig. 2.8.   Scalar illumination.

The flux intercepted by the surface of the sphere will be the same flux as would have been intercepted by a circle of the same radius facing the source (Fig. 2.8). The ratio of the surface area of a sphere ($4\pi r^2$) to the area of a diametral plane (πr^2) is 4, so the ratio of their average illuminations will be 0·25:1.

The illumination on a circular element at P facing the source in Fig. 2.6 will be equal to the illumination vector $\overrightarrow{\text{E}}$. The scalar illumination E_s at P is therefore

$$E_s = 0\cdot25\,\overrightarrow{\text{E}} = \frac{I(\phi)}{4d^2} = \frac{I(\phi)\cos^2\phi}{4h^2}. \tag{2.13}$$

Values of $0\cdot25\cos^2\phi$ are tabulated in the Appendix (Table A.2.II).

The contour lines in Fig. 2.9 are drawn through points on the floor which have the same value of horizontal illumination. Such charts are known as isolux diagrams, iso-illumination diagrams or,

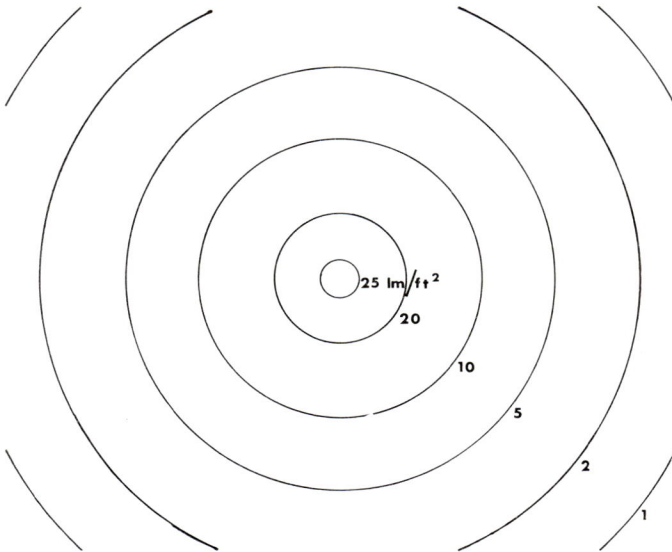

Fig. 2.9.   Isolux diagram 10 feet below 3-foot diameter dome light. For polar curve, see Fig. 2.3.

to avoid linguistic hybrids, as equilux or isophot diagrams [2.3]. They may be based on calculation, using the cosine-cubed rule, or on direct measurement by an illumination photometer. Isolux lines generally ignore light reflected from the room surfaces; only direct illumination is plotted. Figure 2.9 refers to a source ten feet above the ground, and is drawn to a scale of eight feet to one inch. When the source is raised or lowered the isolux contours retain their shape, but their positions and values will change. Thus if the height increases to 15 feet the illumination levels should be reduced in the ratio $(10/15)^2$. The contour lines need not be redrawn, but the scale of the chart is now $(8 \times 15/10)$ feet to one inch, i.e., 12 feet to one inch. The same technique can be applied to asymmetrical isolux lines, and to plots of scalar illumination.

If several similar sources contribute to the illumination at a point the isolux curves can be drawn on tracing paper and several charts, each centred on the appropriate source, can be superimposed on a plan of the floor. The resultant isolux curves are obtained by adding the contribution from each source. A quicker technique is to invert the geometry by centring a single isolux diagram over the test point and reading the illumination at the points corresponding to each source [2.4].

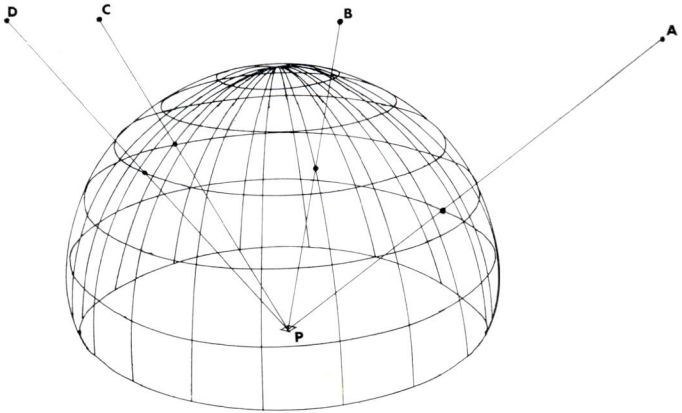

Fig. 2.10.   The unit hemisphere.

2.4. The Unit Hemisphere and the Orthographic Projection

In studying the spatial distribution of the light impinging upon any point P on a horizontal plane it is convenient to visualise the point surrounded by a hemisphere of unit radius, as shown in Fig. 2.10. Circles of altitude and azimuth (analogous to latitude and longitude) define the direction from P of each point on the hemisphere and hence of any point in space inside or outside the hemisphere. Figure A.2 in the Appendix shows a plan view of the unit hemisphere, known as an *orthographic projection*. The radius of its bounding

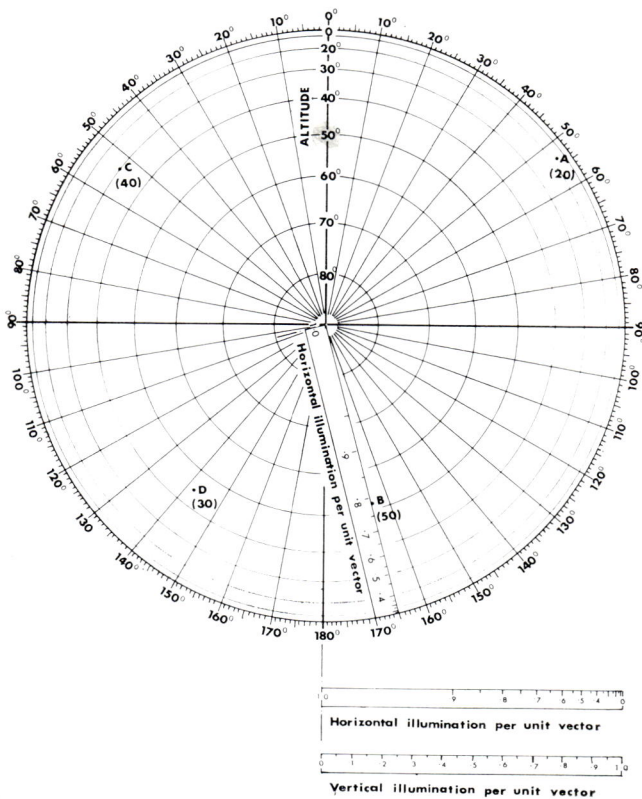

Fig. 2.11. Horizontal illumination at P, due to B, 0·78 × 50 = 39 lm/ft².

circle is the same as that of the hemisphere, i.e., unit radius. On this hemisphere, or on the orthographic projection, we can mark the direction, in terms of azimuth and altitude, of a number of point sources of light, and also the direct illumination vector which each will produce at P, as shown in Fig. 2.11.

The scalar illumination E_s at P will be equal to one-quarter of the sum of the vector components (eqn. (2.13)). The horizontal illumination E_H at P due to each source is equal to $\overrightarrow{E} \sin \alpha$. Sin α is a function of the radial position of the source on the orthographic projection of the unit hemisphere. A radial scale for horizontal

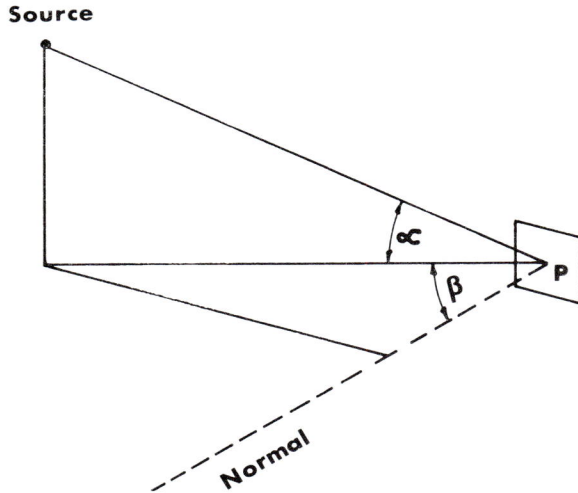

Fig. 2.12.   Illumination on a vertical plane element.

illumination is printed below the orthographic projection in Fig. A.2. To find the horizontal illumination at P we multiply the illumination vector from each source by its distance on the radial scale (Fig. 2.11); E_H is equal to the sum of these products.

The illumination on an elementary vertical surface at P is equal to $\overrightarrow{E} \cos \alpha \cos \beta$, where β is the azimuth angle (i.e., the angle measured in plan) between the direction of the source and the normal to the surface (*see* Fig. 2.12). But $\cos \alpha \cos \beta$ is equal to the length of the perpendicular drawn on the orthographic projection of the unit sphere from the source on to the plane of the element (*see* Fig. 2.13). The component of the illumination vector in any direction in the horizontal plane, such as AP in Fig. 2.13, can therefore be found by

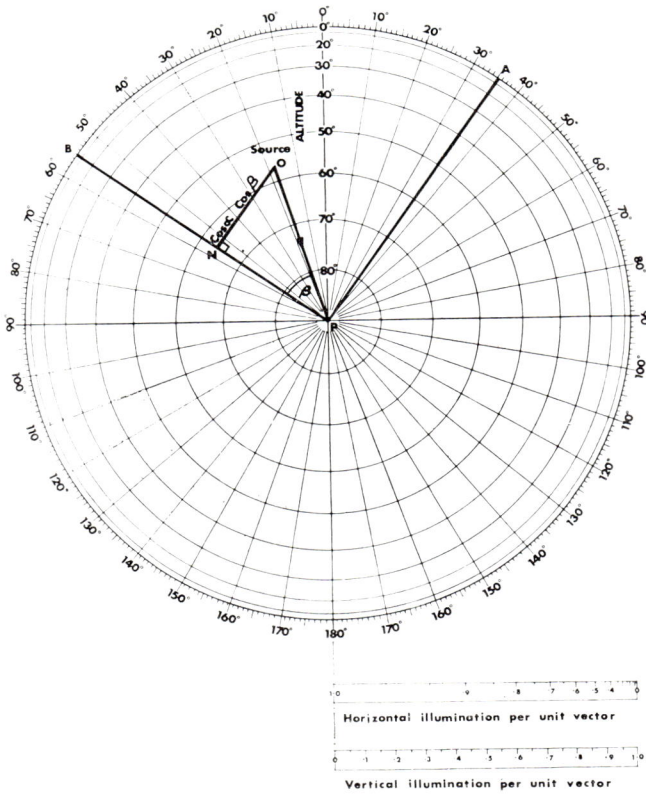

Fig. 2.13.   Illumination on vertical plane at $P = \overrightarrow{E} \cos \alpha \cos \beta$.

measuring the length of the perpendicular from each source on to PB and multiplying it by the corresponding value of $\overrightarrow{E}$. A scale for vertical illumination is printed for this purpose below the orthographic projection in Fig. A.2. Since the illumination vector is the difference between two true illuminations the perpendiculars from opposite sides of PB must have opposite signs—positive and negative. The component of the illumination vector in the direction AP, due to all the sources, is obtained by adding the individual components algebraically.

The component of the illumination vector in the horizontal plane in the direction PB can be found in a similar way by dropping perpendiculars on to PA.

This procedure is analogous to that for finding the moments of forces about a horizontal axis. If each of the sources on the orthographic projection is regarded as a concentrated weight numerically equal to its illumination vector the direction in azimuth of the resultant illumination vector will be the direction of the centre of gravity of the weights. The magnitude of the the horizontal component of the resultant illumination vector will be equal to the sum of the "weights" times the distance of their centre of gravity from the centre of the orthographic projection [2.5].

The vertical component of the vector, i.e., the horizontal illumination E_H at P, can be found by using the radial scale for horizontal illumination. If any of the sources are below the base of the hemisphere their contribution must be subtracted from E_H to find the vertical component of the illumination vector. The magnitude and direction of the resultant illumination vector are found by combining horizontal and vertical components vectorially.

Chapter 3

The Large Light Source

3.1. Luminance

The performance of a large light source, for which the inverse square law fails, cannot be defined by a polar curve of its intensity. Instead it must be broken into elements each of which may have a different intensity distribution. The concentration of flux emitted, reflected or transmitted into a given direction from unit projected area of each element is known as its *luminance, L*.

Figure 3.1 shows a small homogeneous luminous element of area A. The scalar illumination E_s at P will be proportional to the

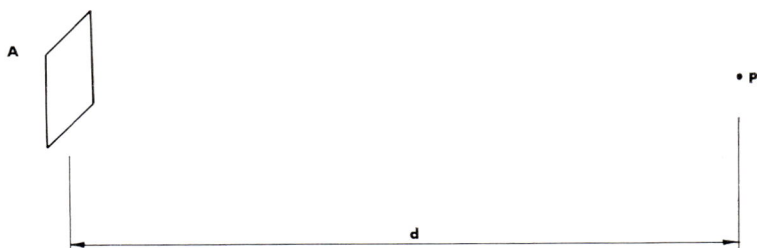

Fig. 3.1.   Illumination due to plane element.

area A or, more generally, to the *projected* area. E_s will also be inversely proportional to the square of the distance of P from the element. We may therefore write

$$E_s = \frac{KA}{d^2} = K\omega \qquad (3.1)$$

where ω is the solid angle subtended at P by the element, and K must be proportional to the luminance of the element.

Equation (3.1) suggests a simple definition for a unit of luminance, the *foot-lambert*. Imagine the point P surrounded by a uniformly luminous sphere (Fig. 3.2). The luminance of the sphere is one foot-lambert if it produces a scalar illumination at P of one lumen per square foot.

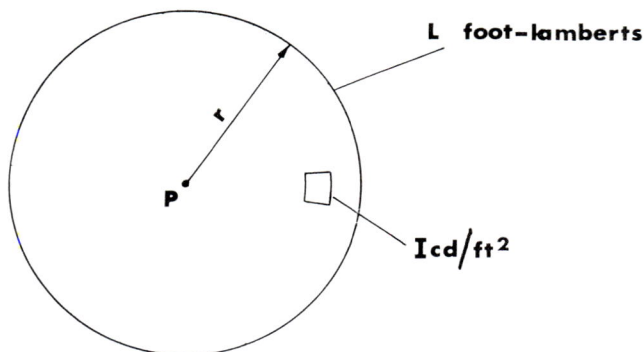

Fig. 3.2.   The units of luminance.

If each element of the sphere has an intensity of I candelas per square foot the scalar illumination E_s at P will be

$$E_s = \frac{I \times \text{area of sphere}}{4r^2} = \pi I.$$

Since the scalar illumination is equal to the luminance L of the sphere in foot-lamberts, we may write

$$L = \pi I. \tag{3.2}$$

This suggests an alternative unit to the foot-lambert. Luminance may be expressed in candelas per unit subtended area. Clearly one foot-lambert equals $1/\pi$ candelas per square foot, or $1/144\pi$ candelas per square inch. The metric system has augmented the prolific family of units of luminance. A table of conversion factors is included in the Appendix (Table A.3).

From the definition of the foot-lambert we may evaluate K in eqn. (3.1). When the light source is a sphere of luminance L foot-lamberts

$$E_s = L \text{ and } \omega = 4\pi$$

$$K = \frac{L}{4\pi}$$

and eqn. (3.1) becomes

$$E_s = \frac{\omega L}{4\pi} \tag{3.3}$$

This can be combined with eqn. (2.13) to express the illumination vector $\overrightarrow{E}$ from a small light source in terms of the source luminance L:

$$\overrightarrow{E} = 4E_s = \frac{\omega L}{\pi}. \tag{3.4}$$

(Note that although eqn. (3.3) is true for a light source of any size eqn. (3.4) applies only to a small light source, since eqn. (2.13) is true only for small sources.)

Equations (3.3) and (3.4) show that the illumination produced at a point by each element of a light source depends solely upon its luminance and upon the solid angle which it subtends. It is *independent of the distance of the source*. For example the illumination produced by the sky seen through a window is independent of the "distance" of the sky; it is completely defined by the direction and luminance of each element of the sky, and on the solid angle it subtends. We can specify the lighting at any point in space solely in terms of the field of luminance surrounding the point.

In a scale model of a building angular subtenses and directional relationships should be the same as in the full-size original. This means that the illumination at any spot inside a scale model will be identical with the illumination inside the original building provided that the correct luminances have been reproduced. The scale of the model is immaterial, but colours and textures must all be accurate. The magnitude and the directional distribution of the light source luminance must also be correct. Fortunately this is easily achieved in the study of daylight, for the same sky illuminates both the model and the real building.

Considerations of geometrical optics show that the illumination on the film in a simple camera will be proportional to the luminance of the object focused on the film. Similarly the stimulation of retinal receptors in the human eye will depend upon the luminance of the object observed. This explains why, despite the inverse square law, objects close at hand look no brighter than objects further away. They subtend larger solid angles, but their luminance, in a perfectly clear atmosphere, is independent of their distance. However, as we saw in Chapter 1, the perceived brightness of an object does not depend solely on its luminance. It does so only when vision is deliberately impoverished, as when viewing portions of objects through small apertures so that they are perceived only as patches of light each seen against a common background. Under these restricted conditions patches of equal luminance should appear equally bright to a normal observer.

3.2. Further Properties of the Unit Hemisphere

Figure 3.3 shows a hemisphere of unit radius constructed about a point P on a horizontal plane. S is a large uniform diffuser of arbi-

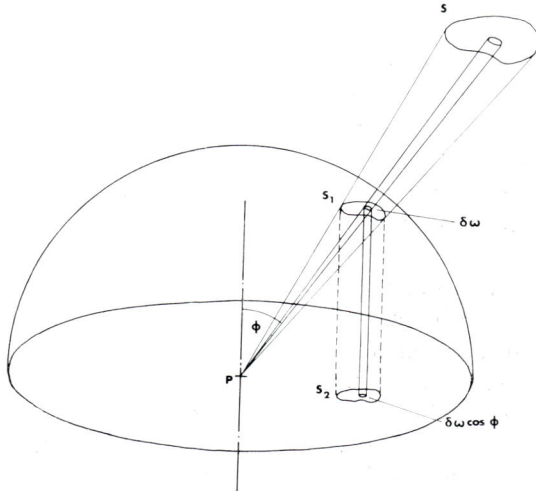

Fig. 3.3.   The unit hemisphere.

trary shape, having a uniform luminance of L foot-lamberts, and subtending a solid angle ω at P.

The scalar illumination E_s at P (eqn. (3.3)) is

$$E_s = \frac{\omega L}{4\pi}.$$

Since ω is equal to the projected area S_1, on the surface of the unit hemisphere, of the original source S *the scalar illumination E_s is directly proportional to the projected area of the source S on the unit hemisphere.*

The illumination vector $\overrightarrow{\delta E}$ at P (eqn. 3.4) due to a small element of S subtending a solid angle $\delta \omega$ at P is

$$\overrightarrow{\delta E} = \frac{L \delta \omega}{\pi}.$$

The horizontal illumination E_H due to this element is

$$\delta E_H = \overrightarrow{\delta E} \cos \phi = \frac{L \cos \phi \, \delta \omega}{\pi}. \tag{3.5}$$

($\cos \phi \, \delta \omega$) is equal to the area intercepted on the unit hemisphere, projected on to the base of the hemisphere. The horizontal illumination E_H due to the whole source S is

$$E_H = \frac{L}{\pi} \times S_2. \tag{3.6}$$

The horizontal illumination is directly proportional to the projected area of S_1 on the base of the unit hemisphere. The direction of the illumination vector at P will be along the line joining the centroid of the surface S_1 to the point P.

If P is exposed to an unobstructed hemisphere of sky of uniform luminance L foot-lamberts, the projected area S_2 on the base of the unit hemisphere will be equal to π. Substituting this value in eqn. (3.6) we find that *the horizontal illumination in lumens per square foot is equal to the sky luminance in foot-lamberts.* Since the distance of the source is immaterial an infinitely extended horizontal luminous ceiling, of uniform luminance L foot-lamberts, would also produce a horizontal illumination of L lumens per square foot at floor level, neglecting reflected light. Since all the flux emitted downwards by the ceiling must eventually strike the floor, the ceiling must, to comply with the principle of conservation of energy, be emitting the same flux as strikes the floor, namely L lumens per square foot. It follows that a uniform diffuser (i.e., a surface whose luminance is the same in all directions) of luminance L foot-lamberts must emit a total flux of L lumens per square foot of surface area. This suggests an alternative definition of the foot-lambert as *the luminance of a uniform diffuser which emits one lumen per square foot of surface area.*

The geometrical properties of the orthographic projection of the unit hemisphere outlined above, together with those described in Chapter 2, make the grid shown in Fig. A.2 in the Appendix a useful conceptual tool for analysing the lighting at a chosen point indoors. Once its general properties are clear it can readily be used to compare the relative contributions of different light sources. It is not customary

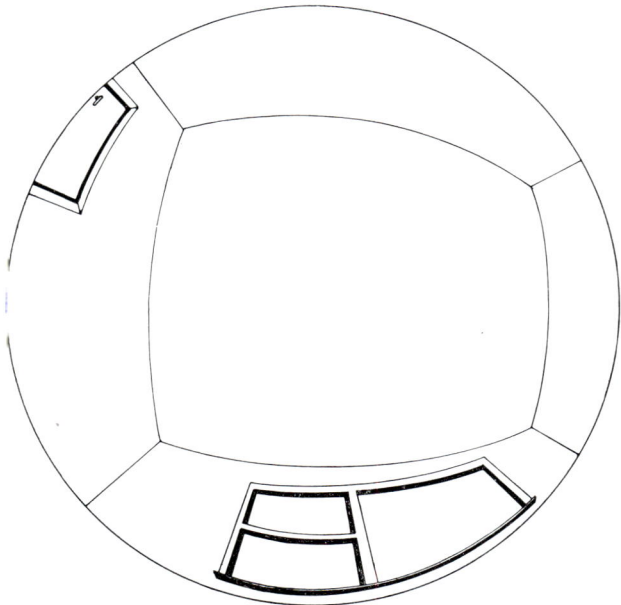

Fig. 3.4. Orthographic view of an interior.

in everyday practice to construct a complete orthographic projection of the walls, ceiling, windows, etc., as in Fig. 3.4, but should this be necessary the following rules will bring speedy results.

All vertical lines in the environment should be plotted radially as lines of constant bearing (azimuth). Their relative directions will be the same as on a plan of the room. All other straight lines in the environment should be plotted as ellipses whose major axis is equal to the diameter of the unit hemisphere. The length of the minor axis of the ellipse corresponding to a given horizontal line will depend upon the ratio of h, the height of the line above the reference plane, to d the shortest distance in plan from the line to the reference point. Figure A.3 shows a family of guide-lines for various values of h/d. It is a simple matter to trace off the appropriate ellipse for a window-head or ceiling-line [3.1].

3.3. Configuration Factors and Sky Factors

The geometrical relation between the unit hemisphere, the surface S and the point P in Fig. 3.3 is conveniently expressed in terms of the configuration factor, C_{ps} (note the order of the subscripts) [3.2]. This is the ratio of the illumination at P, in lumens per square foot, to the luminance of S, in foot-lamberts. As we have seen, this configuration factor is equal to the projected area S_2 on the base of the unit hemisphere.

The sky factor at P is defined as the configuration factor of the patch of sky visible from P, e.g., through a window. This concept is of little practical value, since the sky's luminance is never uniform. However, since the sky factor can be calculated to a high order of accuracy it is still used for settling legal disputes over the interception of daylight by adjacent buildings [3.3]. In such cases the outline of the area of sky visible from a chosen point indoors can conveniently be plotted on the orthographic grid, Fig. A.2, with the help of the guide-lines in Fig. A.3. The sky factor will be equal to the area of the sky expressed as a ratio (or, more commonly, as a percentage) related to the area of the complete circle [3.4].

The configuration factor is a valuable tool for calculating how much light is received by reflection from the walls of a room. In Fig.

3.5 ABCD is a vertical wall of uniform luminance L foot-lamberts, and P is a point on the floor. The horizontal illumination E_H at P will be

$$E_H = LC_{PW} \qquad (3.7)$$

where the subscript W denotes the wall.

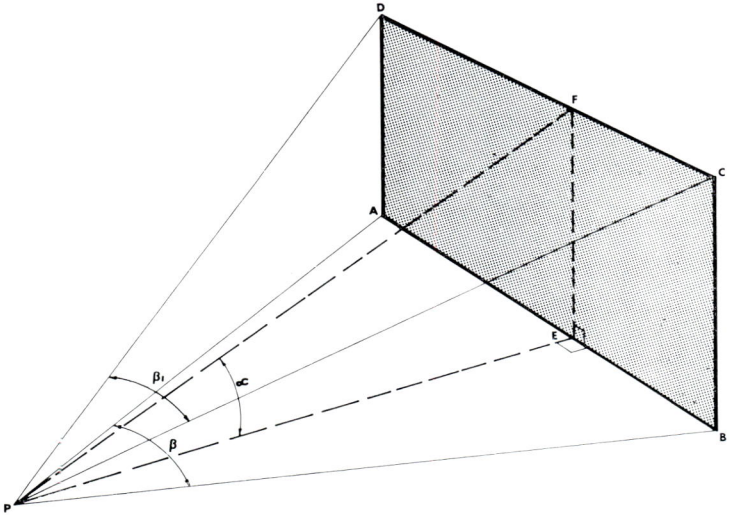

Fig. 3.5.   Configuration factor of vertical wall.

C_{PW} can be found by plotting the projected outline of the wall on the orthographic grid and measuring its area with a planimeter. Alternatively one can express the configuration factor in terms of the angles marked in Fig. 3.5.

Fig. 3.6 shows the orthographic projection of the wall. The area ABCD will be equal to the sector OCD of the ellipse subtracted from the sector OAB of the circle.

$$\text{Area of OCD} = \frac{OF}{OE} \times \text{sector OGH} = \frac{\beta_1 \cos \alpha}{2}$$

$$\text{Area of OAB} = \frac{\beta}{2}$$

Projected area ABCD $= \frac{1}{2}(\beta - \beta_1 \cos \alpha)$

$$C_{PW} = \frac{\beta - \beta_1 \cos \alpha}{2\pi}. \qquad (3.8)$$

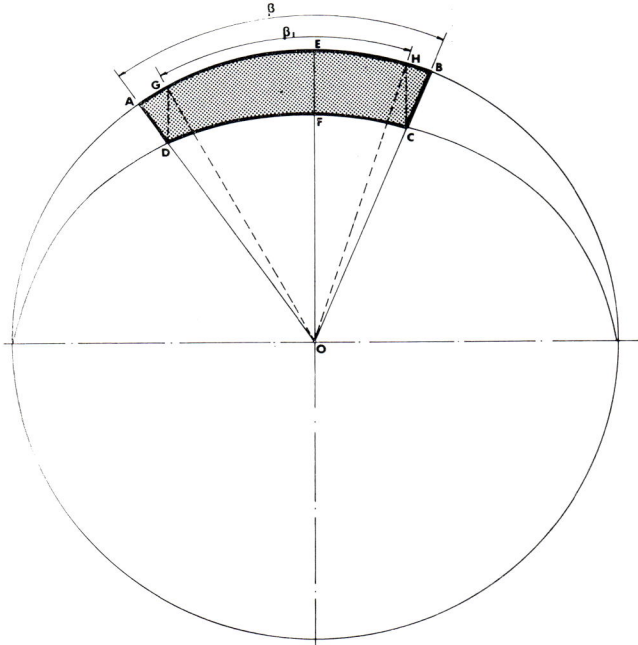

Fig. 3.6.   Orthographic projection of vertical wall.

Substituting this configuration factor in eqn. (3.7) we can find what illumination the wall produces.

Figure 3.7 shows two small flat luminous surfaces of area A_1 and A_2 and uniform luminance L_1 and L_2 foot-lamberts. According to eqn. (3.2) the intensity normal to the surface of A_1 is $L_1 A_1/\pi$ candelas. Since the intensity of the source towards any given direction is proportional to its projected area viewed from that direction we may write

Intensity of A_1 in direction of $A_2 = \dfrac{L_1 A_1 \cos \theta_1}{\pi}.$

Illumination E_2 at A_2 due to light from $A_1 = \dfrac{L_1 A_1 \cos \theta_1 \cos \theta_2}{\pi d^2}$.

Flux F_2 reaching A_2 from $A_1 = E_2 A_2 = \dfrac{L_1 A_1 A_2 \cos \theta_1 \cos \theta_2}{\pi d^2}$.

Similarly flux F_1 reaching A_1 from $A_2 = \dfrac{L_2 A_1 A_2 \cos \theta_1 \cos \theta_2}{\pi d^2}$.

$$\frac{F_1}{F_2} = \frac{L_2}{L_1}. \tag{3·9}$$

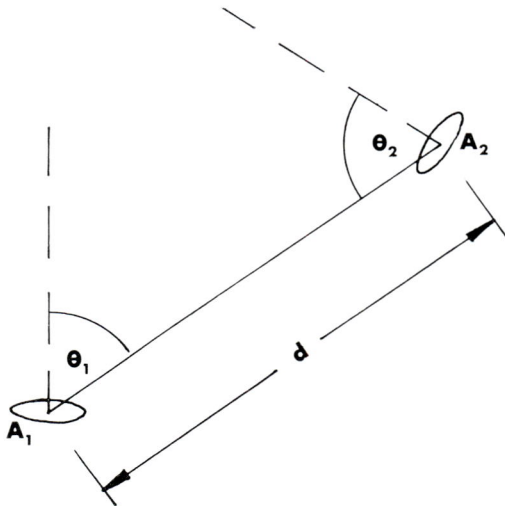

Fig. 3.7.   The law of reciprocity.

Equation (3.9) is known as the *Law of Reciprocity* (not to be confused with the law of reciprocity governing the chemical action of radiation, *see* Chapter 11). Since every object may be resolved into an infinite number of flat elements, each obeying the law of reciprocity, this law must apply to all diffuse surfaces whatever their shape and size. *If any two diffusing surfaces of equal luminance illuminate each other each will receive the same flux from the other.* If their luminances are unequal the flux received by each surface will be proportional to the luminance of the other.

Returning to the situation illustrated in Fig. 3.3, let the luminances of the surface S and the horizontal element P be L_s and L_p foot-lamberts. Let the flux travelling from S to P be F_{sp}, and the flux reaching S from P be F_{ps}. According to the reciprocity law,

$$\frac{F_{sp}}{F_{ps}} = \frac{L_s}{L_p}. \qquad (3.10)$$

The configuration factor C_{ps} was defined as

$$C_{ps} = \frac{E_p}{L_s} = \frac{F_{sp}}{A_p L_s}$$

where A_p is the area of the small element at P.

Substituting the values of F_{sp} from eqn. (3.10),

$$C_{ps} = \frac{F_{ps}}{A_p L_p} = \frac{F_{ps}}{F_p} \qquad (3.11)$$

where F_p = total flux emitted by P.

Equation (3.11) provides an alternative definition of the configuration factor C_{ps} as the proportion of flux, emitted by a flat diffusing element at P, which strikes the surface S.

3.4. The Illumination Field

At each point in a lighted room there must exist an illumination vector; this was proved in Chapter 2. The direction of this vector at any point shows which way a plane element at that point must face to ensure that the difference of illumination between its front and back surfaces is a maximum. The magnitude of the vector is equal to this maximum difference of illumination.

The illumination field in a room may be represented by a family of curves known as flow-lines, the direction of which at every point coincides with the direction of the illumination vector. It is important to distinguish between rays of light, which can and do cross each other, and flow-lines, which cannot. Only one flow-line can pass through any point; and there will be no flow-line in regions where the illumination vector has zero magnitude. The curvature of the flow-lines, surprising at first sight, does not contradict the fact that "light travels in straight lines", any more than Faraday's lines of force were a denial of the laws of electric or magnetic attraction.

It is convenient to replace the infinite number of flow-lines in the illumination field by a finite number of *tubes of flux*, the sides of which are formed by flow-lines. Such a tube of flux is illustrated in Fig. 3.8. It has been shown (eqn. (2.10)) that the difference between

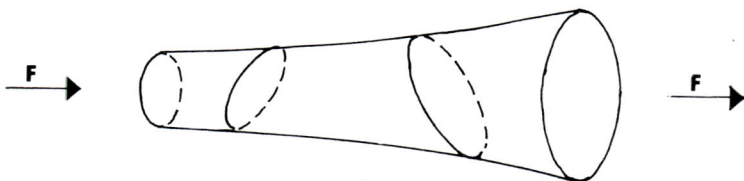

Fig. 3.8.   A tube of flux.

the illumination on the back and the illumination on the front of any element in an illumination field is equal to $(\vec{E} \cos \theta)$ where θ is the angle of incidence at which the illumination vector strikes the element. Every element of the wall of a tube of flux must be parallel to the illumination vector; here θ is equal to ninety degrees and $\cos \theta$ is equal to zero. Thus at every point along a tube of flux the illumination on the interior of the wall must be equal to the illumination on the exterior. The flux passing inwards through any part of the side surface of the tube will be precisely equal to the flux passing outwards.

The net flux incident on every cross-section across a tube of flux is therefore constant; the difference between the flux striking one side and the flux striking the opposite side of each section across the tube in Fig. 3.8 will be the same. The average illumination vector along the whole length of the tube will be inversely proportional to the cross-sectional area of the tube. Since each tube "contains" a constant flux it is convenient to divide the flow-lines so that every tube embraces the same net flux. A diagram of the illumination field in a room could then serve two purposes; besides showing the direction of the illumination vector everywhere it would also reveal the magnitude of the illumination vector, which would be inversely proportional to the density of the tubes of flux. Unfortunately the illumination field is generally three-dimensional, so it is seldom possible to plot such a diagram on a flat sheet of paper.

The illumination field remains a useful conceptual tool. Consider, for example, the field generated by an infinitely long luminous strip of uniform luminance L foot-lamberts [3.5]. Figure 3.9 shows a

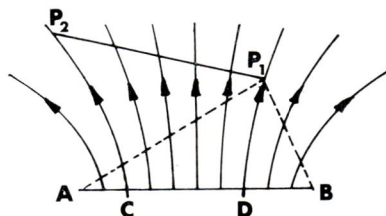

Fig. 3.9. Section through illumination field due to infinite luminous strip.

section AB through such a strip which extends vertically to infinity above and below the sheet of paper. The illumination field will be the same in every plane normal to the major axis of the strip; it can therefore be drawn on a sheet of paper.

At any point P_1 facing the strip the direction of the illumination vector must bisect the angle AP_1B, so each point on a flowline must satisfy the condition that the tangent at every point P along its length bisects the angle APB. This condition is met by the hyperbola through P whose foci are at A and B. Thus the flow-lines in any plane parallel to the sheet of paper form a family of confocal hyperbolas. If such hyperbolas are drawn at equal intervals across AB the same flux will be enclosed between each pair of adjacent hyperbolas.

The line P_1P_2 in Fig. 3.9 represents a section through another infinitely long strip perpendicular to the plane of the paper. The flux intercepted by this strip must be equal to the flux leaving the section CD. The average illumination E (lumens per square foot) on the strip P_1P_2 will therefore be

$$E = \frac{CD}{P_1P_2} \times L \text{ lumens per square foot.} \qquad (3.12)$$

If P_1P_2 is rotated so that light from AB strikes both sides of P_1P_2 eqn. (3.12) will give the difference of illumination between the two sides.

Since a hyperbola is the locus of points the difference of whose distance from the two foci is constant we may write

$$AD - BD = AP_1 - BP_1$$

$$AC - BC = AP_2 - BP_2.$$

By subtraction

$$AD - AC + BC - BD = AP_1 + BP_2 - AP_2 - BP_1$$

$$2CD = AP_1 + BP_2 - AP_2 - BP_1.$$

Equation (3.12) may be rewritten

$$E = \frac{(AP_1 + BP_2 - AP_2 - BP_1)}{2P_1P_2} \times L. \qquad (3.13)$$

Using this equation we may calculate the average illumination produced by one infinitely long strip on another strip without needing to plot the illumination field.

3.5. Transmission and Reflection Factors

The *transmission factor* or transmittance, T, of a translucent material is equal to the light flux which the surface transmits expressed as a ratio, or a percentage, in relation to the flux which falls on the material.

The *reflection factor* or *reflectance*, ρ, is equal to the proportion of the incident flux which is reflected. The *absorption factor*, α, is the proportion absorbed within the material. T, ρ and α all depend upon the nature of the material and upon the direction and wavelength composition of the incident light. The whole of the incident flux must be reflected, transmitted and/or absorbed, so we may write

$$T + \rho + \alpha = 1.$$

If L is the luminance of a light source, the luminance of its mirrored image will be equal to ρL, where the reflection factor of the mirror is ρ. If the same source is viewed through a flat transparent sheet of glass whose transmission factor is T its effective luminance is reduced to TL.

Few reflecting surfaces behave like a mirror. Generally they scatter the light which falls on them, and their luminance is proportional not to the luminance of the light source but to the illumination E which it produces at the surface. In such cases the luminance L_s of the surface, expressed in foot-lamberts, is

$$L_s = \beta E, \tag{3.14}$$

where β is known as the luminance factor, and generally varies with the angle of incidence and also with the direction of viewing. A glossy surface has even more complex reflecting properties, for it scatters some of the reflected light, but acts like a mirror to the remainder. The luminance of a glossy surface therefore depends both on the luminance of the source and on the illumination which it produces. On the other hand the luminance of a uniform diffuser is relatively easy to predict; it is the same for every viewing direction, and its value, in foot-lamberts, is equal to the lumens reflected per square foot of surface. The luminance factor of a uniform diffuser is therefore equal to its reflection factor, ρ, and eqn. (3.14) becomes

$$L_s = \rho E. \tag{3.15}$$

Strictly speaking the uniform diffuser is a theoretical abstraction, for the luminance of real surfaces is never quite independent of the viewing direction. However, eqn. (3.15) is a very convenient expression, and is approximately true for many building materials, so for calculation purposes it is customary to assume that the walls, ceiling and floor of a room are all uniform diffusers.

Chapter 4

Vision

4.1. The Lightness Scale

In Chapter 1 we saw that the only quantitative visual judgement an observer can make accurately is whether two patches of light look equally bright.

   In Chapters 2 and 3 we have seen how the results of this judgement can provide a practical basis for measuring lighting. It must, however, be emphasised that all photometric measurements and calculations are essentially concerned with physical quantities—luminance, illumination or intensity. They do not claim to measure the subjective impression which these physical quantities induce in a human observer. Despite the evidence of the lightmeter we do not feel that the illumination in a room at night is really doubled when we switch on a second light. The fundamental laws of illumination enumerated in Chapter 2, the laws of additivity, transitivity and distributivity, the inverse square law and the cosine law, apply to the radiant stimulus (because light flux is defined in terms of radiant energy). They do not necessarily apply to the physiological or psychological response.

   Our perception of brightness is influenced for example by our emotional needs. Food appears brighter to a man when he is hungry than when he has just eaten. Thus when he is asked to adjust the brightness of a picture of some food to match the brightness of other pictures he chooses a lower luminance for the food when hungry than when he is well fed [4.1].

   The effects of lightness constancy were mentioned in Chapter 1. An experimenter can arrange a series of grey papers in rank order of increasing whiteness or decreasing blackness. If sufficient samples are available he can choose ten papers in subjectively equal steps of greyness ranging from black to white. The best known white/black series of this type is the Munsell "Value" Scale in which greys are numbered from 0 (perfect black) to 10 (perfect white). Figure 4.1

shows an experimentally determined relationship between the
Munsell Value and the reflection factor of matt grey surfaces [4.2].
Two simple approximate formulae have been proposed for finding

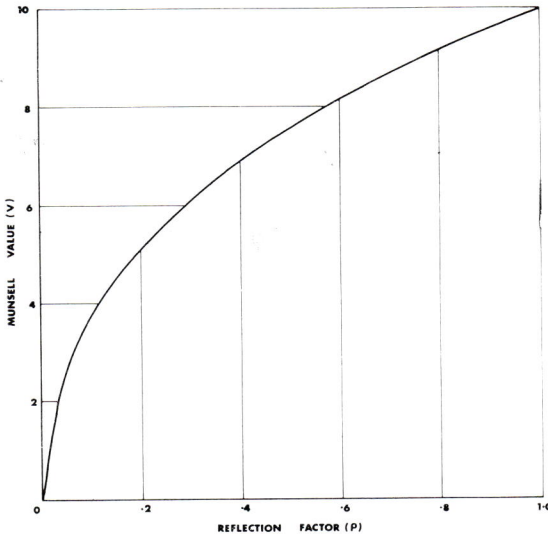

Fig. 4.1.   The Lightness Function.

the nearest Munsell Value for a surface of known reflection factor
or the reflection factor for a given Munsell Value [4.3]:

$$\rho \simeq V(V - 1)$$
$$V \simeq 0 \cdot 5 + \rho^{\frac{1}{2}}$$

where ρ = reflection factor (per cent),

V = Munsell Value.

Since lightness constancy never works perfectly we would expect
the reflectance/value function to depend on such factors as the
illumination of the grey surface, its angular subtense and its background. The curve plotted in Fig. 4.1 was obtained with fairly small

samples seen under high illumination against a light background [4.4]. Seen against a dark background grey patches look lighter, against a light background they look darker; this is the well-known phenomenon of *simultaneous contrast*. However, in a very fine light-and-dark pattern, such as the closely packed dots in a half-tone

Fig. 4.2.   Each grey patch has the same reflection factor.

reproduction, this effect is reversed; the contrast is reduced rather than enhanced. Each of the grey patches in Fig. 4.2 has the same reflection factor, and this illustration shows that the contrast effect is by no means a simple matter. The lightness of the grey squares will appear to vary a little with the distance from which the page is viewed and the angle through which it is tilted.

Thanks to the phenomenon of lightness constancy a small reduction in illumination makes little difference to the appearance of a lighted room, but a large reduction (say a 50:1 reduction) causes a significant breakdown of constancy. The apparent contrast between greys and whites increases, while the apparent contrast between greys and blacks decreases. This is known as the *white dog effect* (because white dogs are conspicuous in the dark—an example of the phenomenon), and can be observed by studying Fig. 4.2 in a darkened room. One consequence of the white dog effect is that when a room is darkened subtle gradations of modelling seem harsher. Reduced illumination makes all colours look less vivid; in Munsell terminology their chroma seems to decrease. The warm colours— those containing red or orange—lose their chroma more readily than do the cooler colours, and so a darker room tends to look colder; this phenomenon is associated with a change in the shape of the relative luminous efficiency ($V(\lambda)$) curve at low luminances, making the eye more sensitive to blue than to red light wavelengths.

These four effects of the failure of constancy in dim lighting— the blackening of greys, the harsher modelling, the loss of vividness and the impression of coldness—are summed up by the term *gloom*. It would be a mistake to suppose that an impression of gloom in a daylit room is caused just by low illumination. Where constancy most commonly breaks down is on the window wall, which receives no direct illumination from the window but which is seen right next to a bright area of sky. The excessive contrast is responsible as much as the low illumination for the gloom which results.

To relieve gloom it is necessary to restore lightness constancy. The most obvious way of achieving this is by raising the illumination over those surfaces where constancy has failed. Much can be accomplished however by careful choice of surface finish. White surfaces retain constancy well; this is the positive side of the white dog effect. Even limited white areas, such as a white skirting-board or white trim round a doorway, will tend to impart constancy to adjacent surfaces. Vivid colours retain constancy better than do muted ones, and can be safely used on surfaces normally in shadow. Organic materials having a distinctive texture also tend to preserve constancy, while glossy finishes have the opposite effect. The coldness associated with dim lighting can be counteracted by using "bronze" rather than "grey" glass for windows with low transmission properties, and, in artificial lighting, by choosing the warmer fluorescent lamp colours.

4.2. Weber's Law

When a surface A of luminance L_A is viewed against a background of luminance L_B the contrast C between the surface and its background is defined mathematically as

$$C = \left| \frac{L_A - L_B}{L_B} \right|. \tag{4.1}$$

If both surface A and background B are good diffusers and receive the same illumination the luminances will be proportional to their respective reflection factors; in this case

$$C = \left| \frac{\rho_A - \rho_B}{\rho_B} \right|. \tag{4.2}$$

The phenomenon of lightness constancy makes this a particularly useful way of measuring the contrast C, because the subjective greyness of A and B will be unaffected by changes of illumination over the range within which lightness constancy prevails. Within this range equal values of C will correspond to roughly equal noticeable degrees of observed contrast between surface and background. If we write ΔL for $L_A - L_B$, eqn. (4.1) becomes $C = \Delta L / L$;

i.e., equal *proportional* differences of luminance are equally noticeable.

This relationship, sometimes known as *Weber's Law*, holds roughly for other senses too. It applies, for instance, to intervals of loudness and pressure and to estimates of the heaviness of weights.

Weber's Law, like lightness constancy, seldom holds precisely in practice. Thus if Weber's Law were true, equal ratios of reflection factor in Fig. 4.1 would always correspond to equal steps of lightness. This is roughly so, over much of the scale, and Weber's Law can be said to offer a handy first approximation to the judgement of a human observer under good interior lighting conditions.

Weber's Law is most easily studied in the laboratory under "threshold" conditions, i.e., conditions where the contrast is barely perceptible. Figure 4.3 shows typical values for the contrast of small

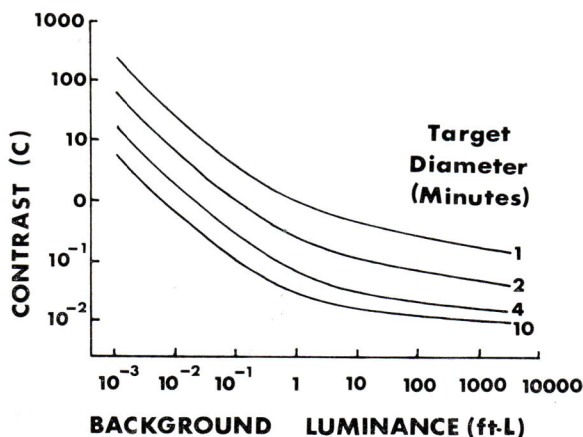

Fig. 4.3.   Visual discrimination for 0·2 second exposure time and 50 per cent accuracy of detection.

targets which had a 50/50 chance of detection against a given background luminance when exposed to view for 0·2 second [4.5]. If Weber's Law applied precisely, each of the curves would be a horizontal straight line, i.e., the minimum perceptible contrast would be the same for every value of luminance. This is seen to be roughly true for well-lit interiors where luminances generally exceed 1 foot-lambert.

In judging the natural lighting in a room lit by side windows we are influenced not only by the illumination indoors but also by how the interior illumination compares with the illumination which prevails outside. When the outdoor illumination changes, then, other things being equal, the indoor illumination will change proportionately, and the contrast C between indoor and outdoor luminances will remain the same. Bearing Weber's Law in mind it is reasonable to express the illumination at a point indoors as a percentage of the

horizontal illumination prevailing at that moment under an unobstructed hemisphere of sky outside. This percentage is known as the
daylight factor.

The daylight factor has a practical advantage over the illumination as a guide to the amount of light a window will provide at a
point indoors, since the daylight factor may stay reasonably constant throughout wide fluctuations in sky illumination. This is
strictly true, however, only when the *pattern* of sky luminance is
static. When the sun shines or when isolated clouds are about the
indoor illumination will no longer be a constant fraction of the outdoor illumination. The use of daylight factors is therefore restricted
in practice to solidly overcast weather. In temperate climates this is
a prudent basis for natural lighting design, but in dry sunny regions
where overcast conditions are uncommon daylighting is expressed
in terms of the illumination rather than the daylight factor.

Though Fig. 4.3 was obtained under threshold conditions our
everyday experience confirms that Weber's Law is roughly valid for
higher contrasts too. Thus the visibility of black print on white paper
is reduced only a little when the paper is in shadow, so that the
value of C defined by eqn. (4.2) offers a rough-and-ready indication
of contrast for the general run of visual tasks in well-lit interiors.
Nevertheless Weber's Law, like lightness constancy, does break down,
especially when luminance is low, and indeed the failure of Weber's
Law under poor lighting conditions explains why official "codes" of
recommended illumination have come to be published for various
activities.

4.3. Illumination and Visual Performance

At first sight one might suppose that the illumination needed for a
given visual task could be obtained by interpolating among the curves
in Fig. 4.3. Consider, for example, a tiny black detail ($\rho_A = 0\cdot04$)
on a sheet of white paper ($\rho_B = 0\cdot8$). From eqn. (4.2), $C = 0\cdot95$. If
the luminance of the paper is $0\cdot1$ foot-lambert (corresponding to an
illumination of only $0\cdot125$ lm/ft^2) the detail will be just detectable if
it subtends 2 minutes of arc, i.e., 7 thousandths of an inch at a reading
distance of 1 foot. Clearly this example tells us something about

human visual performance, but very little about the illumination required for accurate vision; we seldom need to detect 7-thou details, and when we do we use a magnifying glass! Contrast thresholds such as those plotted in Fig. 4.3 refer to a 50/50 chance of correct detection when observers know exactly what to watch for, and where and when it will appear. Unfortunately there is as yet no sure way of deriving, from such laboratory data, the appropriate illumination for a given practical visual task. A sophisticated technique evolved by Blackwell provides a broad theoretical basis for the illumination levels currently prescribed by the Illuminating Engineering Society, New York. Blackwell maintains that practical visual tasks in industry may be treated as laboratory discrimination tasks whose contrast has been reduced by a "field factor" of the order of 6:1 or 15:1, depending upon the laboratory data selected [4.6]. Though there is some experimental support for the "field factor" concept it is by no means certain how much the experimentally determined contrasts should be reduced, and it is clear from Fig. 4.3 that at luminances greater than one foot-lambert a small change in contrast can have a vast effect on the illumination specified—indeed an infinite effect when Weber's Law holds. Blackwell's approach would be extremely difficult to validate rigorously, and is not generally accepted outside North America, but it has stimulated a considerable number of laboratory investigations which should eventually clear up the outstanding points of controversy.

Various lighting authorities have published lighting recommendations for specific visual tasks in terms of the illumination on a horizontal working plane. This is the plane on which the work is normally done, e.g., the desk top in an office. Unless otherwise specified the working plane is assumed to be 2 ft 9 in above floor level. The recommended illumination levels are not founded upon fundamental visual performance data, except in a very general sense. The illuminations specified differ among themselves and have tended to increase from decade to decade. This does not seriously detract from their validity, however, for they are essentially codes of good current engineering practice. Good lighting practice cannot be fixed and codified for all time; it can be expected to vary from country to country, and within each country to respond to technical advances, to economic changes and to rising expectations.

Various workers have set out to establish a correlation between the prevailing illumination and the speed and accuracy with which a given repetitive visual task can be carried out. In the best known studies in this field, by Weston [4.7], observers were required to mark in pencil a gap in a printed geometrical pattern. Though Weston's results have been shown [4.8] to be consistent with Blackwell's they are open to the criticism that subjects became more skilled in the course of the experiment, and that the effect of practice can outweigh the effect of changing illumination [4.9]. Hewitt reports a study of the practice effect on a Weston-type visual task [4.10] in which a similar session of work was repeated every day for several weeks; performance improved considerably over the first 5 days and significantly during the next 7 days.

The obvious recourse—to measure the effect on production in an actual factory when a new lighting scheme giving a higher illumination is installed—is also beset with difficulties. Many reports, mostly of an anecdotal nature, have been published; not unnaturally a dramatic increase in productivity is more likely to be publicised than a negative outcome. The most serious study in this field, which opened a new phase in industrial sociology, was carried out by Mayo, in the Hawthorne Works of the Western Electric Company, Chicago, during the nineteen-twenties. Two groups of employees were chosen, a control group who worked under the same lighting through the experiment, and a test group for whom the illumination was increased. As had been anticipated, the output in the latter group improved, but, unexpectedly, the output from the control group also increased. The investigators then reduced the illumination for the test group, whose output increased once more. Subsequent experiments on a smaller group of workers, extending over a period of five years, studied the effects of piece-work, work-breaks, and a shorter working week, and finally reverted to the conditions at the start of the experiment, when output broke all previous records. Details of the investigation may be found in text books of social psychology [4.11]; the chief conclusion was that production depended much more on the workers' satisfaction with each other and with their job than on their physical environment. Where better lighting leads to higher output this is more likely to be due to improved morale than to better agreement with Weber's Law.

The Hawthorne experiment also has methodological implications. The very fact that investigators seem to be taking an interest in a person may raise his self esteem and hence his attitude to the job he is doing, and this "Hawthorne effect" is the bugbear of any experiment which seeks to elucidate the influence of lighting on performance or productivity outside the artificial environment of the laboratory.

4.4. Scalar Illumination as a Lighting Criterion

There is no doubt that a bright working environment induces a sense of well-being, and most people, given the choice, would prefer an illumination higher than can be proved necessary for carrying out ordinary visual tasks. As lighting standards rise and Weber's Law applies more and more closely it is becoming less necessary to justify illumination recommendations by reference to what is needed for a prescribed level of visual performance.

In judging how bright a room looks, people are influenced as much by the appearance of vertical surfaces as by the horizontal illumination. Indeed in rooms devoted to social contact and recreation the conventional horizontal working plane has little meaning. A better criterion of the subjective adequacy of the illumination in such an interior is the scalar illumination at eye level. In some cases the horizontal illumination is positively misleading. It tends to exaggerate the effectiveness of overhead lighting, and conceals the important effect of light-coloured floors or working surfaces on the overall impression which interior lighting creates. This distinction between horizontal illumination and scalar illumination is particularly important at the back of side-lit rooms where light from the sky strikes horizontal surfaces very obliquely. Here a lightmeter which registers the illumination on a horizontal plane may well suggest that the light is insufficient, even though the scalar illumination, indicating the overall subjective impression, is adequate [4.12].

Daylight factors are usually expressed in terms of either horizontal or scalar illumination, and it should be noted that the scalar daylight factor at a given point is equal to the scalar illumination at that point, expressed as a percentage of the simultaneous *horizontal* (not scalar) illumination under an unobstructed hemisphere of sky.

The choice between scalar and horizontal illumination as a basis for specifying daylight factors in a given interior will depend upon the principal purpose of the natural lighting. Where there is work to be carried out on a horizontal plane, the daylight factor should generally be based on the horizontal illumination; where the work is not paramount, but a bright ambience is required, the scalar illumination is the more appropriate criterion. In many cases it is wise to consider both horizontal and scalar daylight factors.

4.5. Modelling

Horizontal and scalar illumination are at best an index of quantity. The appearance of a given three-dimensional object will depend also upon the directional qualities of the lighting, i.e., on the distribution of luminance on all points visible from the object. The orientation of

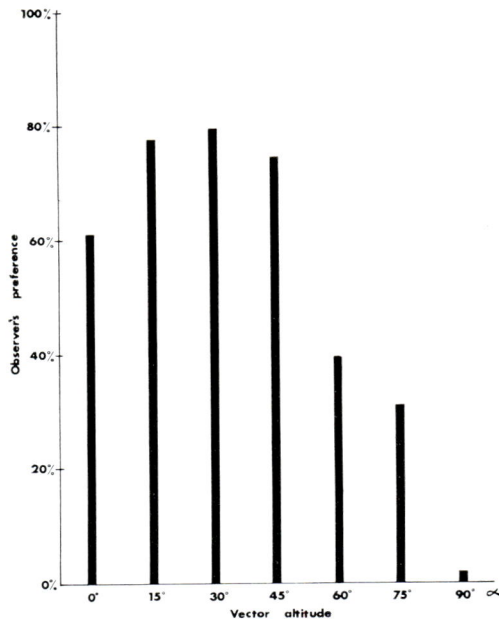

Fig. 4.4.   Preferences for vector direction.

the illumination vector at a point in space shows the subjectively dominant direction of the lighting at that point. Most people prefer the lighting to have some directional character, but all directions are not equally popular. The histogram in Fig. 4.4 shows the statistical breakdown of a choice by "single-transferable-vote" in which observers were presented with a random sequence of vector directions and asked to rank them in order of preference. No clear-cut favourite emerged but it is clear that most people prefer the predominant direction of the lighting to be from a direction between 15 and 45 degrees above the horizontal; lighting from directly overhead has been found least acceptable. Although the direction of the illumination vector is an important guide to the quality of the modelling the magnitude of the vector, i.e., the maximum difference of illumination across opposite faces of an object, does not necessarily tell us how intensely the object will appear to be modelled. Weber's

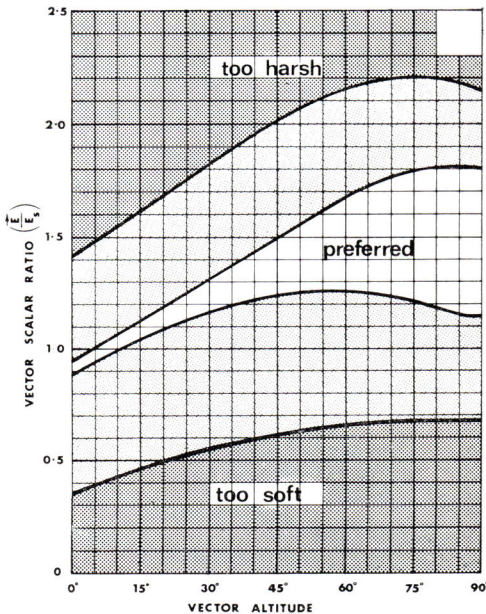

Fig. 4.5. Subjective impressions of the modelling of the human features, expressed in terms of the vector direction and the vector/scalar ratio, $\vec{E}/\bar{E}_s$.

Law reminds us that the contrast depends not on the illumination vector but on the ratio of the illumination difference to the average illumination. This ratio, i.e., the illumination vector $\overrightarrow{E}$ divided by the scalar illumination E_s, is known as the *vector/scalar ratio* $(\overrightarrow{E}/E_s)$.

When the vector/scalar ratio is high the modelling will appear harsh. When the ratio is low the modelling will seem soft. Generally people prefer slightly higher vector/scalar ratios as the predominant direction of the lighting approaches the vertical (*see* Fig. 4.5). The vector/scalar ratio is a valid index of modelling only when Weber's Law more-or-less prevails. When the lighting is dim the white dog effect makes objects seem more harshly modelled.

Chapter 5

Daylight and Sunlight

5.1. Natural Light

An obvious characteristic of daylight is the way in which it varies from season to season and even from minute to minute. Figures 5.1 (a) and 5.1 (b) show typical recordings of horizontal illumination

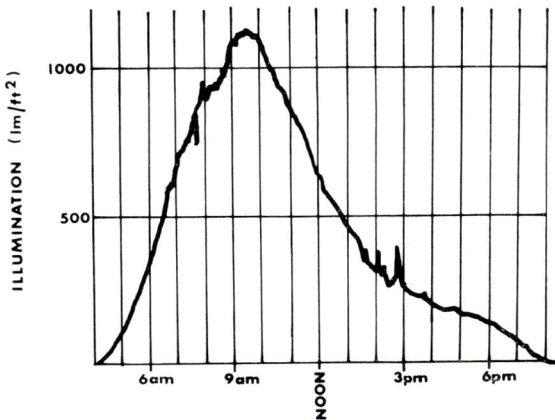

Fig. 5.1 (a).   Horizontal illumination on photocell exposed to east octant of sky, but protected from direct sunlight, on a sunny day.

on a photocell exposed to one quarter of the sky vault and screened from direct sunlight [5.1]. Figure 5.1 (a) shows the illumination for the east octant on a clear sunny day; wisps of cloud caused the small irregularities. Figure 5.1 (b) was obtained from the west octant on an overcast day.

This characteristic variability—the "living" quality of daylight, as it is sometimes called—is not necessarily a disadvantage. Under some circumstances a varying illumination may assist concentration, as measured by span of attention. It does mean, however, that daylight prediction cannot usefully aim at a high degree of accuracy, and that design criteria have inevitably to be based on a statistical treatment of meteorological data.

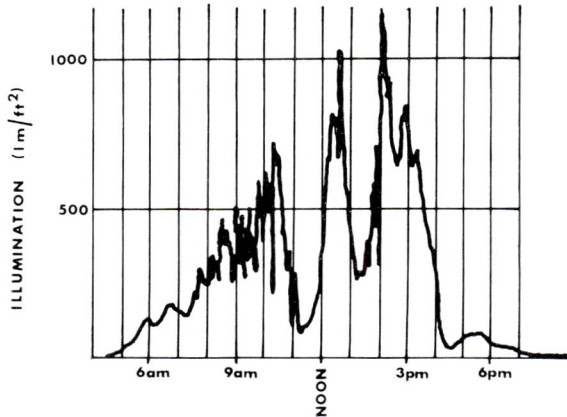

Fig. 5.1 (b).   Horizontal illumination on photocell exposed to west octant of sky, but protected from direct sunlight, on an overcast day.

Figure 5.2 shows typical curves of radiant flux, E_λ, plotted as a function of the wavelength λ. The upper curve relates to direct light from the sun at an altitude of 30 degrees above the horizon [5.2]. The middle curve represents "total daylight", i.e., radiation from the whole sky vault including direct sunlight [5.3]; in 1967 the Commission Internationale de l'Éclairage (C.I.E.) adopted this curve as a basis for defining a standard light source for laboratory purposes, to be known as C.I.E. Standard Illuminant D_{6500}. The lower curve shows Standard Illuminant C, accepted as an approximation by the C.I.E. in 1931; most published data on glazing materials are still based on Illuminant C, but this is likely to be replaced soon by Illuminant D_{6500} which is more typical of true daylight, especially in the shorter wavelengths.

5.2. Sunlight

The apparent movement of the sun across the sky is due to the rotation of the earth about the sun and about its own axis, but the rise and

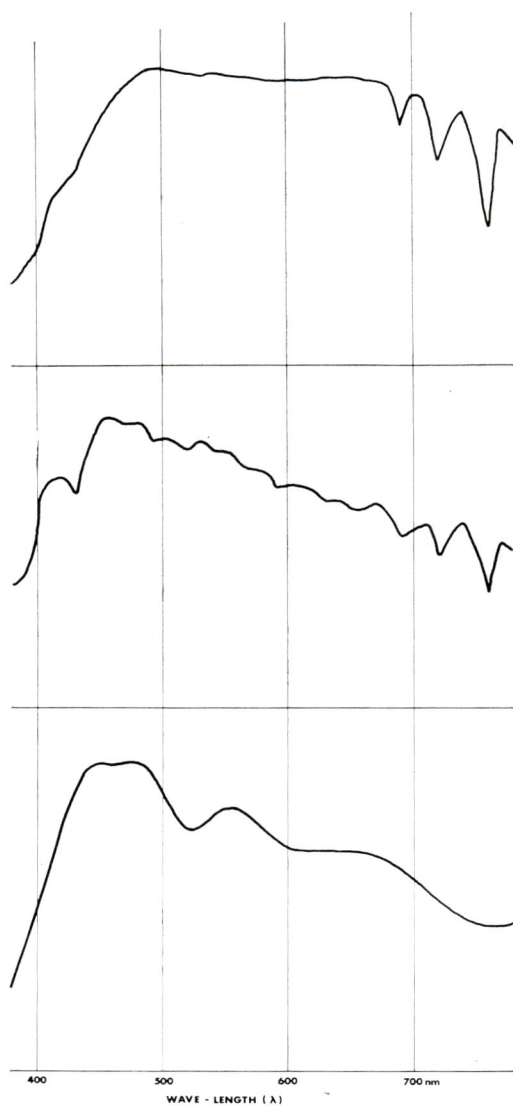

WAVE - LENGTH (λ)

Fig. 5.2. Daylight spectral power distributions. Top curve: direct sunlight. Middle curve: total daylight (C.I.E. Illuminant D_{6500}). Bottom curve: C.I.E. Illuminant C.

fall of daylight at a point on the earth's surface is easier to follow if we revert to the primitive notion of a flat static earth and a circulating sun. The sun's apparent orbit on any day of the year can then be represented on an orthographic projection of the unit hemisphere. Figures A.5.1 to A.5.7 in the Appendix each shows a set of sunpaths for one chosen day in each of the twelve months. A separate orbit is drawn for each latitude, at intervals of 10 degrees. These are drawn to the same scale as used for Fig. A.2, so radial distances are still

Table 5.1

Equation of Time

Date	To obtain clock time
January 21	Add 11 min to solar time
February 23	Add 13½ min to solar time
March 21	Add 7½ min to solar time
April 16	No correction required
May 21	Subtract 3½ min from solar time
June 22	Add 1½ min to solar time
July 24	Add 6½ min to solar time
August 28	Add 1½ min to solar time
September 23	Subtract 7½ min from solar time
October 20	Subtract 15 min from solar time
November 22	Subtract 14 min from solar time
December 22	Subtract 1½ min from solar time

proportional to cos α, where α is the altitude, above the horizon, of any point on the unit hemisphere. One can estimate the altitude of the sun by measuring its distance from the centre of the chart (i.e., the centre of the circle, not the centre of the ellipse); the corresponding distance from the left end of the upper radial scale in Fig. A.5.8 shows the height of the sun above the horizon.

To avoid cluttering the diagrams the sunpath lines are discontinued where they intersect; the sun's direction at sunrise and sunset are not shown. Vertical lines, known as *hour lines*, indicate the moment at which the sun will reach various points on its orbit. The time indicated by the hour lines is actually *solar time*, i.e., the time that would be indicated by a correctly oriented sundial. It makes no allowance for daylight saving, so where an hour's daylight saving is in force we must add one hour to each of the times indicated. Also since the earth rotates through 360 degrees in a day, or 1 degree every 4 minutes, we should apply a *longitude correction* of 4 minutes for every degree of longitude measured from the central meridian of the

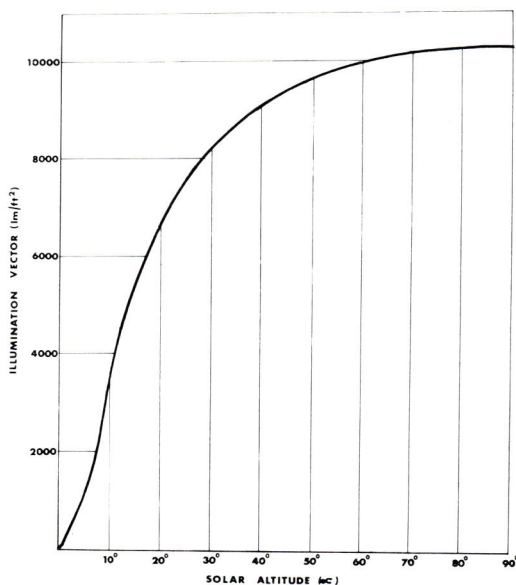

Fig. 5.3.   Illumination vector from direct sunlight.

relevant time zone. For places west of the central meridian we add 4 minutes per degree to the solar time; for every degree east we sub-tract 4 minutes to obtain the "clock time". A further correction, known as the *equation of time*, is due to small asymmetries in the earth's movement about the sun; this is too small to affect lighting design but is included in Table 5.I for completeness.

The formula for converting solar time to clock time is:

Clock time = solar time + daylight saving + equation of time + longitude correction of 4 minutes for every degree west of central meridian.

The illumination vector on the earth's surface due to direct sunlight, excluding light reflected or scattered by the rest of the sky or by the ground, is affected, even on a clear day, by the depth of atmosphere through which the sun has had to penetrate. Figure 5.3 shows how the magnitude of this vector varies with the altitude of the sun, i.e., with its distance from the centre of one of the ortho-graphic sunpath diagrams [5.4]. This enables us to prepare radial scales for the illumination vector due to direct sunlight, and for the

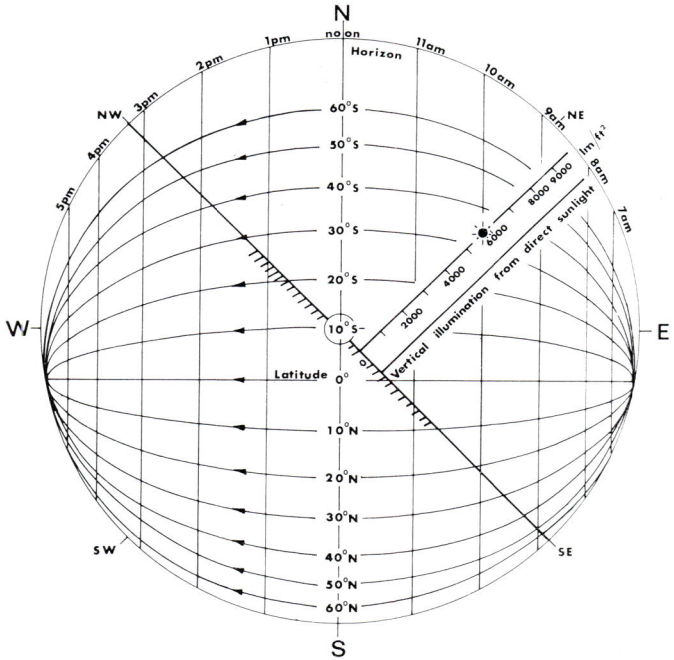

Fig. 5.4. Direct sunlight illumination on vertical wall—5700 lm/ft².

direct solar illumination on a horizontal plane, Fig. A.5.8. These scales read from left to right, like the scales below Fig. A.2. Suppose, for example, that we wish to find the horizontal illumination due to direct sunlight under clear weather conditions at 10 a.m. solar time on February 23rd at Sydney, New South Wales (latitude 34° S). We find the position of the sun on Fig. A.5.3. (for February 23rd) on the 10 a.m. vertical hour-line, interpolating between the intercepts of the 30° S and 40° S sunpath ellipses. The distance of this point from the centre of the chart is measured; the corresponding distance on the radial scale for horizontal illumination from direct sunlight gives the required value of E_H—approximately 5850 lumens per square foot.

Some idea of the sunlight illumination on a vertical wall is obtained by assuming a constant illumination vector of 10,000 lumens per square foot, irrespective of the height of the sun. The procedure illustrated in Fig. 2.13 can then be applied, the vertical illumination being proportional to the normal from the sunpath to a line through the centre of the base of the unit hemisphere parallel to the vertical wall. Suppose, for example, that the wall faces north-east at the same latitude, 34° S. The illumination on the wall at 10 a.m. on February 23rd is found by dropping a perpendicular from the sun's position and measuring its length against the radial scale for vertical illumination, Fig. A.5.8, as shown in Fig. 5.4. Since this scale is linear the orthographic sunpath is essentially a tilted graph of the vertical illumination.

5.3. The Distribution of Sky Luminance

The luminance of the sky on a cloudless day is entirely due to sunlight scattered in its passage through the atmosphere. In outer space, beyond the earth's atmosphere, the empty sky looks black. The luminance of a given element P of sky, viewed from the ground, will depend upon:

The altitude α of the element P above the horizon.

The zenith angle Z_s of the sun, measured in radians from the zenith.

The angle ψ subtended at the ground by P and the centre of the sun.

Kittler has proposed the following formula for the luminance L_p of an element of clear sky [5.5]:

$$L_p = \frac{L_z(1 - \varepsilon^{-0.32\cos\alpha})\,(0.91 + 10\varepsilon^{-3\psi} + 0.45\cos^2\psi)}{0.274(0.91 + 10\varepsilon^{-3Z_s} + 0.45\cos^2 Z_s)} \quad (5.1)$$

where L_z is the sky luminance at the zenith.

Equation (5.1) was adopted by the Commission Internationale

de l'Éclairage (C.I.E.) in 1967 as an agreed definition of a standard clear sky luminance distribution upon which illumination prediction techniques can be based.

In Fig. 5.5 calculated values of clear sky luminance are plotted for a solar altitude of 50 degrees, on an orthographic projection of the unit hemisphere. An obvious consequence of the $\cos^2 \psi$ term in the numerator of eqn. (5.1) is that a dark patch of sky is observed where $\psi = 90°$; this can be clearly seen outdoors on a sunny day.

It would be possible, but very laborious, to estimate the horizontal illumination from a patch of clear sky seen through a window by drawing its outline on an orthographic projection such as Fig. 5.5 which shows luminance curves for a given solar altitude. The area between successive luminance contours could be measured by a planimeter; the horizontal illumination would be equal to the product of the mean luminance and the area on the orthographic projection (*see* Section 3.2 above).

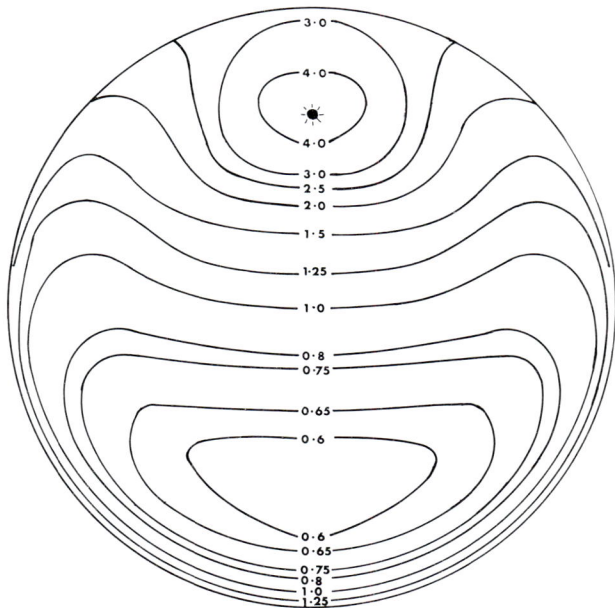

Fig. 5.5.   Luminance distribution of cloudless sky, when sun is 50 degrees above horizon. Zenith luminance taken as unity.

On a completely overcast day no direct sunlight reaches the ground. So thoroughly is the sky light scattered that the pattern of sky luminance is virtually symmetrical about the zenith. The luminance of an overcast sky is lower at the horizon than overhead. A typical gradation of luminance from zenith to horizon is expressed by the equation

$$L_\alpha = \frac{L_z}{3}(1 + 2 \sin \alpha) \qquad (5.2)$$

where L_z = luminance of the sky at the zenith,

$\qquad L_\alpha$ = luminance of the sky at an altitude of α degrees above the horizon.

Equation (5.2) was adopted by the Commission Internationale de l'Éclairage in 1955 to define the C.I.E. Standard Overcast Sky [5.6], which now provides the basis for daylight prediction under overcast sky conditions.

In practice the pattern of sky luminance on an overcast day is affected by the reflection factor of the ground, and this may vary from season to season. Equation (5.2) represents observed conditions fairly closely when the ground is dark. The luminance overhead ($\alpha = 90°$) is then three times the horizon luminance ($\alpha = 0°$). During the winter in mountainous regions of Central and Northern Europe and in much of the U.S.S.R. overcast skies seldom occur except when the ground is covered with snow. Under these conditions the zenith luminance is only about twice the horizon luminance, and Kittler has proposed a modified formula to describe the corresponding sky luminance pattern [5.7]:

$$L_\alpha = \frac{L_z}{2}(1 + \sin \alpha). \qquad (5.3)$$

5.4. Illumination from the Sky Vault

Even on an overcast day the horizontal illumination outdoors is still a function of solar altitude. Krochmann has proposed the following empirical equation to describe this relationship [5.8]:

$$E_H = 30 + 1950 \sin \alpha. \qquad (5.4)$$

Since the direct illumination from the sun, on a sunny day, far exceeds the illumination from the rest of the sky the latter is generally ignored. Alternatively the sky may be treated as a hemisphere of uniform luminance. Its luminance in foot-lamberts would be equal to the illumination in lumens per square foot on a horizontal surface shielded from direct sunlight on a clear day.

Three curves for the horizontal illumination due to the sky are plotted in Fig. 5.6. These show the illumination from a cloudless sky

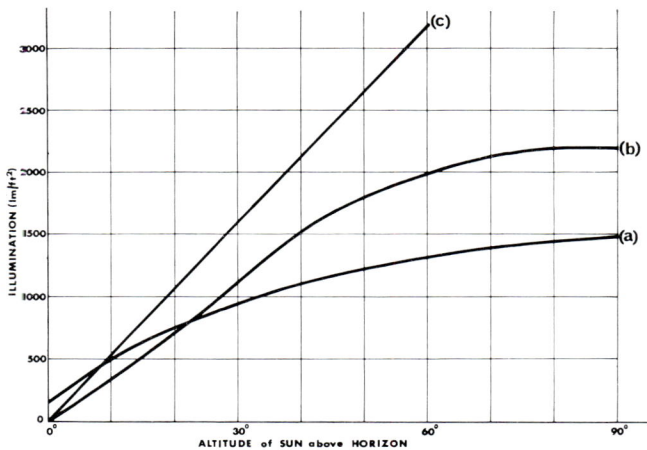

Fig. 5.6. Illumination from sky vault: (a) cloudless day (excluding direct sunlight); (b) overcast sky; (c) mean sky illumination (excluding direct sunlight).

[5.9], the illumination from a completely overcast sky (eqn. (5.4)), and the mean illumination on a photocell exposed to the whole sky in a temperate climate but screened from direct sunlight. At first sight it seems surprising that for most solar altitudes the mean sky illumination exceeds the illumination on cloudless days and on overcast days. This is explained by the fact that the brightest parts of the sky are usually the sunlit edges of white clouds, which would be hidden on a completely overcast day and absent on a cloudless day. The mean sky illumination varies with solar altitude but is almost independent of the degree of cloud cover over a fairly wide range of nebulosity. It can be expressed by the equation [5.10]:

$$E_H = 53\alpha \qquad (5.5)$$

where α is the solar altitude measured in degrees.

The three curves in Fig. 5.6 enable us to construct the three radial scales for sky illumination, Fig. A.5.8. Since natural lighting design in temperate climates is generally based on a sky having the overcast luminance distribution defined by eqn. (5.3) it might seem logical to use eqn. (5.4) for predicting sky illumination. Only in Germany is it customary, however, to assess natural lighting on the basis of the illumination under overcast conditions [5.11]. In hot dry climates where overcast skies are extremely rare natural illumination will depend mainly on direct or reflected sunlight. In temperate climates daylight calculations normally assume the statistical mean values of outdoor illumination from the whole sky excluding direct sunlight. Local meteorological data should be used where available; elsewhere eqn. (5.5) or the radial scale for mean sky illumination should be employed.

Few data are available on the effect of atmospheric pollution on natural illumination. Kimball in 1921 compared the illumination in the clear air of Washington with the illumination in the smokiest district of Chicago where "the fall of soot . . . was so considerable that on some days it was necessary to clean the photometer mirror after each series of observations" [5.12]. During the winter, the season of heaviest pollution, he found the average intensity of sunlight reduced to one-half of that in Washington, and the average sky illumination reduced by about one-third. Thanks to newer sources of fuel and methods of heating such a high degree of pollution would be unusual today even in heavily industrialised regions.

Chapter 6

Instrumentation for Daylight Studies

6.1. Visual Photometry

Instruments for photometry—the measurement of light—are of two types, visual and photoelectric. The former relies on a visual judgement by a human observer, the latter converts light into electricity and measures the radiation in terms of the current or voltage generated.

As was inferred in Chapter 1 an observer's ability to estimate luminance or illumination is very limited. All we can do with confidence, even under favourable conditions, is to judge whether two patches of light look equally bright, or to assess which patch looks brighter. All visual photometry therefore involves adjusting the brightness of one patch of light so as to match another. For best results there should be no visible separation between the two patches.

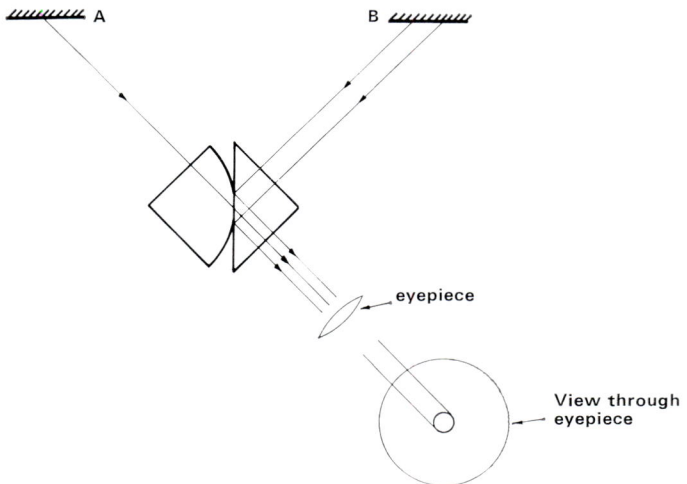

Fig. 6.1.   The Lummer–Brodhun Cube.

One device for achieving this, the Lummer–Brodhun cube, is illustrated in Fig. 6.1. The "cube" comprises a right-angled glass prism whose curved hypotenuse has a flat central area in optical contact with another right-angled prism. Light from a luminous surface A passes straight through the central areas into the eyepiece. It is seen against a background of light from B totally reflected inside the second prism. When A matches B in luminance and in colour the central spot is indistinguishable from its background. If A is brighter than B the luminance of the spot will be seen to exceed the background luminance.

Vision has been found to depend upon small movements of the eye which cause individual retinal receptors to be stimulated in a fluctuating manner. Clearly when the central spot in the Lummer–Brodhun cube disappears small eye movements can no longer provide a fluctuating stimulus at the centre of the field of view, and

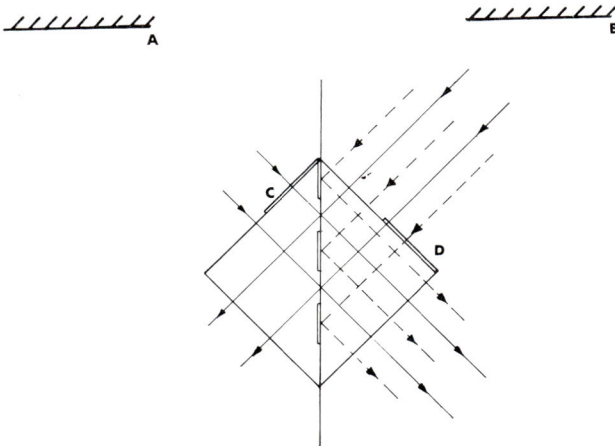

Fig. 6.2.   The Lummer–Brodhun Contrast Cube.

visual discrimination suffers accordingly. A modified arrangement known as the Lummer–Brodhun contrast cube is shown in Fig. 6.2. In the version illustrated both prisms have flat faces, but the left-hand one is etched with the pattern shown hatched in Fig. 6.3. When the two prisms are in optical contact light from A passes straight through both prisms in the region shown unshaded in Fig. 6.3, while rays from B are totally reflected. Thin glass slides at C and D reduce the luminance of the trapezia so that when A and B are equally bright the backgrounds merge and the trapezia, far from disappearing in this type of cube, stand out in equal contrast from their backgrounds.

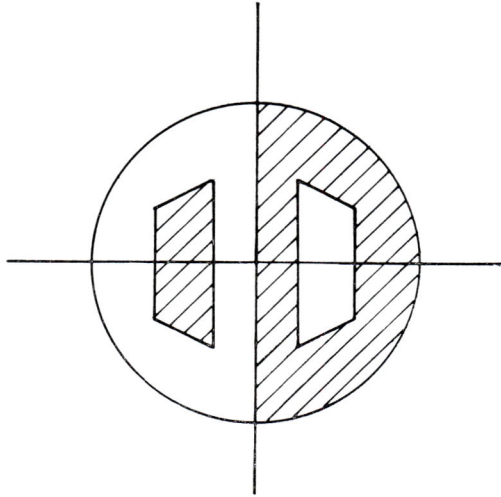

Fig. 6.3.   The Lummer–Brodhun Contrast Field.

The Lummer–Brodhun contrast cube is therefore more sensitive to small luminance differences than is the simple Lummer–Brodhun cube, but both impose a severe constraint on photometric equipment in which they are incorporated; in order to measure the luminance of surface A the apparatus must incorporate a comparison surface B the luminance of which is variable, calibrated, and capable of matching A in brightness and preferably also in colour.

Figure 6.4 shows diagrammatically a portable visual photometer for measuring luminance. Screens for absorbing stray light are omitted for clarity. The luminance of A is found by adjusting the luminance of the opal glass B until a match is obtained. The attenuator may take the form of an annular neutral filter wedge or an adjustable diaphragm. Alternatively the luminance of B may be adjusted by varying the distance of the lamp L or by varying its apparent distance by means of a movable reflecting prism. If A is much brighter than B the range of the instrument can be increased by incorporating a selection of calibrated neutral filters at C. If B is much brighter than A the neutral filters should be inserted at D. Other filters can be incorporated at C for modifying the apparent colour of B.

An important source of error in this type of instrument is the intensity of the lamp L. Its output will be sensitive to small changes in voltage, so its circuit generally incorporates a rheostat and a volt-meter or a milliammeter. Low voltage lamps, such as miners' cap lamps, have the advantage that their filaments are resistant to sudden

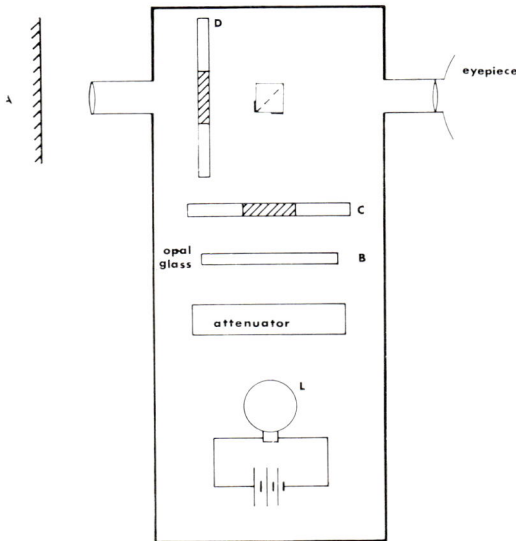

Fig. 6.4.   The elements of a visual luminance meter.

shock; their stability is improved by operating below their rated voltage. Electrical connections should be soldered, and the lamp should have a screw cap to prevent resistance from fluctuating. No readings should be taken until such a lamp has been alight for at least half a minute as the resistance of the filament supports, and the amount of heat they conduct from the filament, change appreciably while they are warming up.

If a visual photometer is used regularly its calibration should be checked and recorded weekly so that slow "drift" in the lamp output can be detected at an early stage. If it is used only occasionally its calibration should be checked before and after each time it is used.

6.2. Photoelectric Cells

Photoelectric photometry relieves the instrument designer of the need to provide a comparison surface to match the luminance under test. It also eliminates the effects of individual variations in sensitivity to light of different wavelengths.

The response of the photocell must be proportional to the incident illumination. Its response for radiation of a given wavelength should therefore be proportional to the relative luminous efficiency $V(\lambda)$ of that wavelength. Its response should also be proportional to the total incident flux over the whole range of conditions in which it is to be used.

There are four types of photocell:

1. *Photoemissive cells and photomultipliers,* in which incident radiation causes electron emission from electrodes inside a transparent envelope. Cells of this type are widely used for precision photometry, but the necessity for amplifier circuits and for elaborate electrostatic screening and temperature control makes them generally unsuitable for the measurement of natural illumination.

2. *Photoconductive cells,* in which the resistance varies with the incident radiation. Cadmium sulphide photoconductive cells have a spectral response covering the visible wavelengths. The resistance variation is not proportional to the incident radiation, though electronic circuits have been devised in which the current change is approximately proportional to illumination. The need for complex circuitry and the sensitivity to temperature variations have discouraged their use in natural lighting studies.

3. *Phototransistors,* which utilise the fact that the current flowing in a semi-conductor such as germanium is a function of the light

falling on it. For daylight measurements their advantage of compact-
ness is outweighed by their sensitivity to changes in temperature and
by the destructive effect of subjecting the device to excessive illumination.

4. *Photovoltaic cells*, also known as barrier-layer or rectifier
cells, which convert incident light directly into electrical energy and
therefore, unlike other types of photocell, do not require an external
source of electricity. Two types are used for photometry, the silicon
cell and the selenium cell. Both have a short-circuit output current
roughly proportional to the incident illumination. Although the
former has a far greater sensitivity, and is available in much smaller
sizes, it is seldom used for photometry because it does not respond
to blue or violet light and because its sensitivity is markedly affected
by ambient temperature especially when it is not operated under
short-circuit conditions.

It is impossible in this book to discuss the characteristics of every
photocell in detail. The following notes relate to the selenium photovoltaic cell which, thanks to its simplicity and robustness, is by far
the commonest choice for daylight photometry. Figure 6.5 illustrates

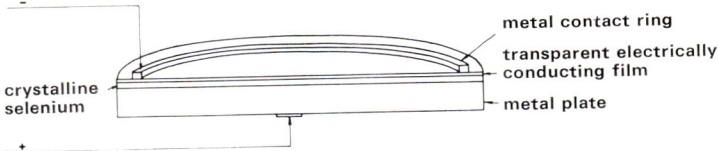

Fig. 6.5.   Section through a selenium photovoltaic cell.

the general construction of the selenium cell. This is normally
clamped into a non-conductive housing whose pressure ensures a
good electrical connection between the upper metal contact and the
conductive window over the selenium and also between the metal
base and the positive contact (usually a phosphor-bronze spring);
the cell must not be clamped too tightly or the thin selenium layer
may become fractured. Soldered electrical contacts would have
obvious advantages over spring contacts but the soldering must be
carried out without damaging the cell by overheating. This can now
be achieved by using low melting-point solder, and a "potted"

selenium photocell with soldered leads is commercially available, mounted behind glass and sealed from moisture and chemical attack.

Figure 6.6 shows the equivalent circuit of a selenium photocell. Since the photovoltaic cell is also photoconductive the forward

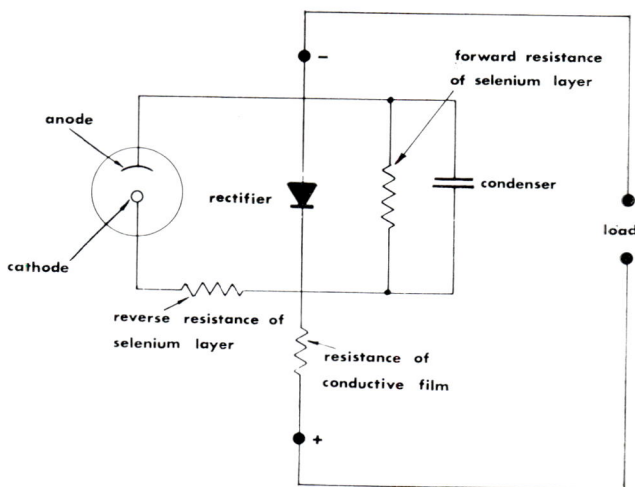

Fig. 6.6.   Equivalent circuit of selenium photovoltaic cell.

resistance of the selenium layer decreases as the illumination rises. Therefore although the current generated in the "barrier-layer" on the selenium surface is proportional to the incident illumination the current flowing in the external circuit will depend also upon the external load resistance. Figure 6.7 shows typical response curves for a 45 mm diameter cell. Since the forward resistance of the selenium is inversely proportional to the area of the cell a large area cell will show increasing departures from linearity; in practice this disadvantage is offset by the higher sensitivity of a large area cell which enables it to be used with a microammeter of lower resistance.

All selenium cells suffer from "drift" effects which vary from cell to cell. Except when in use cells should be stored in the dark. Their response tends to decrease during their first few minutes of exposure to light, so readings should not be taken until they have been illuminated for about ten minutes. "Fatigue" effects occur on prolonged exposure to high illumination. With a high load resistance the cell's sensitivity gradually rises; with a small load it falls. The magnitude of this effect depends on the amount and spectral composition of the incident radiation; after a period in darkness the cell regains its original characteristics.

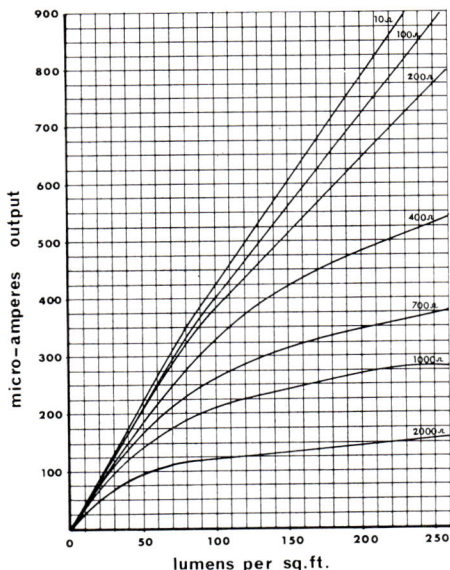

Fig. 6.7.   Response of typical 45 mm dia. selenium cell with various load resistances.

Selenium cells are generally used in series with a galvanometer or microammeter, the choice of which must be a compromise between the conflicting demands of sensitivity which calls for a high coil resistance, and linearity of photocell response which implies a low resistance. For a 45 mm diameter cell, the galvanometer resistance should not exceed about 200 ohms for illuminations up to 50 lumens per square foot [6.1].

This limited load resistance and the high capacitance of the cell make it impossible to use conventional electronic circuits for amplifying the photocell response. Daylight photometry involves the measurement of such a wide range of illuminations that some arrangement for controlling the sensitivity of the photocell/galvano-meter combination is essential. The simplest circuit for this purpose is shown in Fig. 6.8. The shunting resistance R should be approxi-

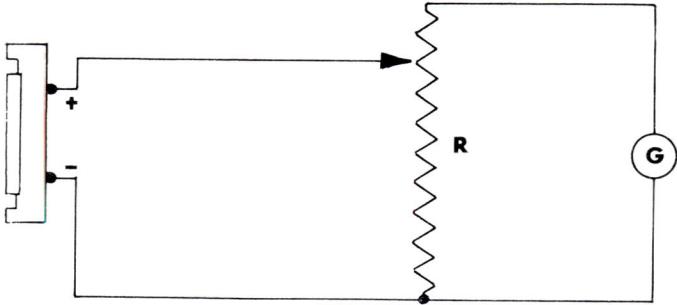

Fig. 6.8.   Sensitivity control for photocell and galvanometer.

mately equal to the critical damping resistance of the galvanometer; a higher resistance produces a sluggish response, while a lower resistance makes the instrument overshoot and flutter. The sensitivity of the circuit is controlled by moving the tapping-point. When the tapping-point is central the load resistance will be maximum; if the load is then excessive the galvanometer is unsuitable for use in this circuit.

The sensitivity of the photocell/galvanometer combination is affected by temperature. Most galvanometers have a temperature coefficient of about 0·4 per cent per degree Celsius. Individual selenium cells differ in their response to temperature; some are unaffected, some increase and some decrease in sensitivity. A temperature coefficient of the order of 0·3 per cent per degree Celsius is typical, but the value tends to increase with illumination and with load resistance.

Both linearity and stability are improved by operating the cell under short-circuit conditions, i.e., such that there is no potential difference across the photocell terminals. In the Campbell–Freeth circuit illustrated in Fig. 6.9 (a) the variable potential V is adjusted to produce a null reading on the galvanometer G [6.2]. The cell is then operating under short-circuit conditions and the current through the fixed resistance R must be equal to the short-circuit photocurrent i.

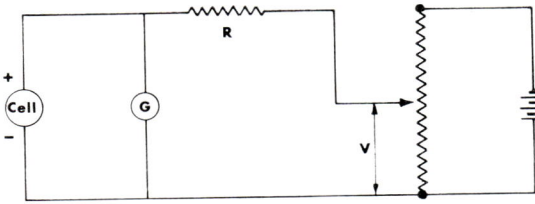

Fig. 6.9 (a).   Campbell–Freeth circuit.

Since $i = V/R$, it is convenient to arrange for R to have a large resistance, of the order of a megohm, and to find i by measuring V with a sensitive d.c. voltmeter. In the modified circuit of Fig. 6.9 (b) the resistance r is of the order of 100 ohms and the range-changing resistance R can have values between 1 and 100 kilohms. The photo-current is proportional to the reading of the milliammeter.

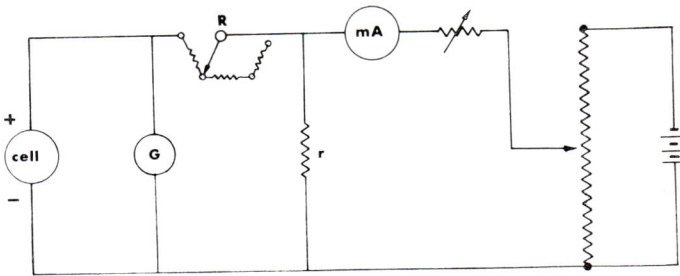

Fig. 6.9 (b).   Modified Campbell–Freeth circuit.

In the alternative circuit illustrated in Fig. 6.10 the potential V is adjusted so that the reading of the microammeter is unaffected by repeatedly opening and closing the press-button switch S. The microammeter, which can be of a sensitive high-resistance type, will then give a direct reading of the short-circuit photocurrent [6.3]. This circuit gains an advantage in portability over the Campbell–Freeth circuit by eliminating one of the meters, but the repeated surges of current cannot be good for the microammeter.

Figure 6.11 shows the relative response of a selenium photocell to radiant energy of different wavelengths. Again significant differences are found between apparently similar cells. It is possible to improve the spectral response of a photocell, so that it approximates to the C.I.E. relative luminous efficiency curve, by placing a greenish filter over the front of the cell. Any improvement will be gained only at the expense of the photocell's sensitivity, and since the spectral

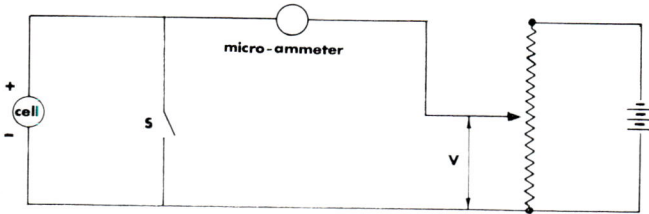

Fig. 6.10.   Single meter blocking circuit.

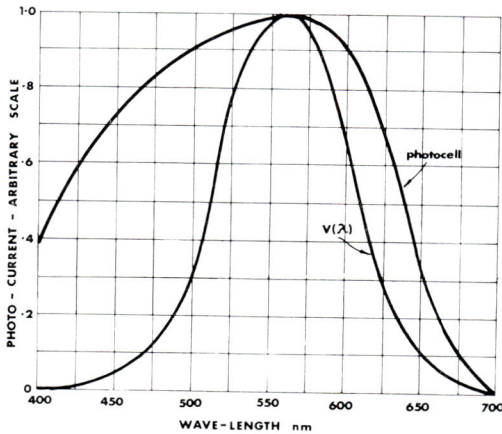

Fig. 6.11.   Spectral response of typical selenium photocell.

transmission of the filter will vary with the direction of the incident light it is impossible to obtain "colour correction" for every direction simultaneously. For daylight photometry it is essential to use a correction filter, albeit imperfect, if only to prevent the cell from responding to infrared and ultraviolet radiation for which the transmission factor of window glass is significantly lower than for visible light. Several firms offer slightly different gelatin filters for this purpose and it is wise to choose the most suitable filter for each separate photocell individually by checking the cell's response to coloured lights of known intensity (*see* Section 6.5).

It will be appreciated from the above notes that although photoelectric photometry avoids the need for visual judgements of equal brightness or equal contrast it has pitfalls of its own, and novices should be on their guard against the blissful credulity that can be induced by a seemingly unambiguous meter reading. Neither a galvanometer nor a photocell lasts for ever. The former may attract magnetic dust or its coil may be jolted into a non-uniform portion of the magnetic field. The latter may be permanently damaged by exposure to damp or to the fumes from certain adhesives (unless "potted"), or by excessive temperature (over 70°C) or excessive illumination. Careless handling, and even repeated tightening and loosening of terminals, may impair electrical contacts within the

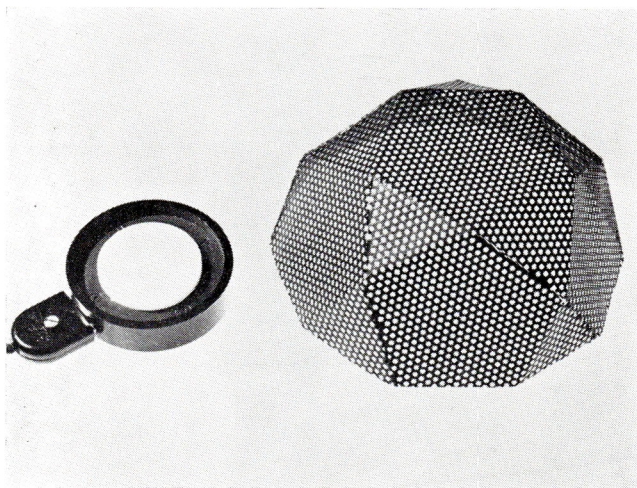

Fig. 6.12.   Left: cosine corrected photocell. Right: polyhedral attenuator.

photocell housing. Fortunately daylight photometry is mainly concerned with ratios of illumination or luminance, so reliable readings can be obtained by observing the following rules punctiliously:

1. Check the null-point of electrical instruments, with the photocell in circuit, before and after each series of readings, by covering the surface of the cell.

2. Check the linearity of the cell and circuit, before and after each series of readings. Figure 6.12 illustrates an easily made attenuator whose transmission factor is almost independent of the colour and direction of incident light. It is cut out of perforated metal to the shape shown in Fig. 6.13, folded into shape, soldered, and blackened internally. Its transmission factor once determined will not change, and will be equal to the ratio of the response of the photocell covered by the attenuator to its response when uncovered so long as the cell and circuit remain linear.

3. Choose a position where the illumination or luminance will remain steady throughout the series of readings. Note the photocurrent at this point and return to check it periodically.

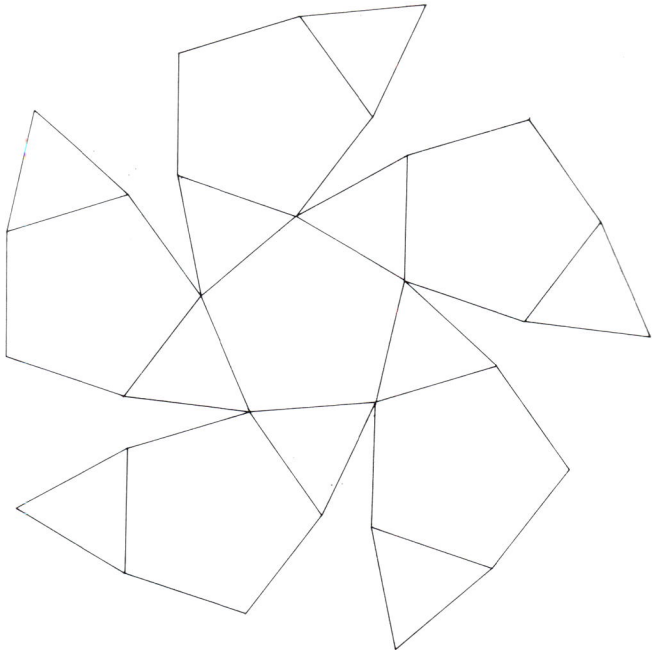

Fig. 6.13.   Shape of attenuator before folding.

Photometric procedures should respect the inherent limitations of available apparatus, which abounds in pitfalls for the unwary. It is a waste of time to devise complex circuits for improving the linearity of a cell which has not been properly colour-corrected, especially as each additional circuit element is a potential source of trouble.

6.3. The Measurement of Illumination

A luminance meter such as that illustrated in Fig. 6.4 can be used for estimating the illumination on a flat surface by measuring the luminance of a diffusing test-plate, of known reflection factor, placed on the surface. But the selenium photocell, on account of its flat sensitive face, is so well adapted for measuring surface illumination that it has virtually superseded all other photometers for this purpose.

If a photocell is to measure illumination accurately it must, when illuminated only by a small source of light, be free from "cosine error", i.e., when the cell is rotated its response must be proportional to the cosine of the angle of incidence of the rays of

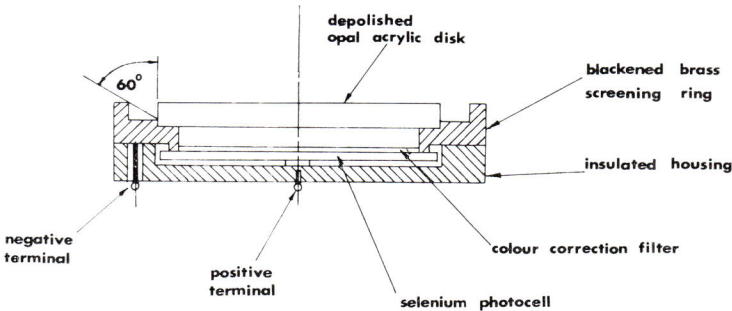

Fig. 6.14.   Section through cosine corrected photocell, after Pleijel and Longmore [6.5].

light. Contrary to a common misconception a bare selenium photo-voltaic cell has very little cosine error. Unfortunately the transmission factor of the glass or gelatin colour-correction filter falls off at oblique incidence. The lip of the housing, which must be robust to ensure good electrical contact with the cathode of the photocell, casts a shadow over the surface of the cell. Both these factors introduce cosine error at high angles of incidence.

Of the various techniques available for correcting the cosine error [6.4] the method devised by Pleijel and Longmore [6.5], illustrated in Fig. 6.14, has proved most useful for daylight photometry as it does not unduly increase the size of the device, reduces its sensitivity by only about 50 per cent, and is unaffected by polarisation of the incident light. "Cosine correction" is achieved by placing a disk of depolished opal acrylic plastics above the cell and surrounding it by a blackened brass ring whose edge must be precisely level with the surface of the disk. Opal acrylic plastics may not be optically neutral—it tends to scatter blue light more than red light—so colour correction should be checked with the disk in position.

The device illustrated in Fig. 6.15 for measuring scalar illumination consists of a table-tennis ball glued to the skirt of a miniature selenium photovoltaic cell. By blackening selected areas of the ball one can make the response of this device almost independent of the

Fig. 6.15.   Device for measuring scalar illumination.

direction of the incident light [6.6]. An alternative technique, sufficiently accurate for most purposes, is to read the illumination on each of the four faces of a regular tetrahedron by means of a cosine-corrected photocell; the scalar illumination is roughly equal to the average of the four readings.

The vector illumination at a point may be found by reading the illumination on the six sides of a cube. The magnitude and direction of the vector resultant is obtained by combining the six measurements vectorially.

Daylight factors inside a model room under an artificial sky are found by measuring the "indoor" illumination and expressing it as a percentage of the "outdoor" illumination measured by the same photocell. Unfortunately it is much harder to use the same technique for measuring daylight factors in a real building. First it is necessary to wait for an overcast sky; one may have to wait several weeks for this even in an inclement climate. If measurements have to be made far from home one must rely upon weather forecasts and still risk a fruitless journey. Even under overcast conditions there is no assurance that the C.I.E. standard luminance pattern will prevail so one cannot hope for readings to be accurately repeatable. Since overcast skies are associated with rain and cold the measurement of outdoor illumination can be uncomfortable for observer as well as bad for instruments. The measurement of natural illumination indoors can also present problems on an overcast day since artificial lighting is likely to be in use, so daylight readings in factories or offices must be crammed into lunch breaks or week-ends when lamps can be switched off. Outdoor illumination fluctuates from moment to moment on an overcast day (see Fig. 5.1 (b)), so indoor and outdoor readings must be taken as nearly simultaneously as possible. This implies the use of two photocells, one indoors and one outside, accompanied by long leads so that the operator can take comparison readings in rapid succession. Alternatively two operators, linked by a "walkie-talkie", can read indoor and outdoor illuminations simultaneously.

If the outdoor cell is to be exposed to an unobstructed sky it must be placed on the roof, assuming (and this is not always so) that the roof is both accessible and unobstructed. Petherbridge and Collins [6.7] have circumvented this difficulty by measuring the luminance of a patch of the overcast sky at an altitude of 42 degrees above the horizon, instead of measuring the outdoor illumination, for this is the altitude at which the luminance in foot-lamberts of the C.I.E. standard overcast sky is numerically equal to the unobstructed out-door illumination in lumens per square foot. The outdoor luminance meter may be placed outside the window (provided, of course, that it opens) and avoids the need for trailing leads from floor to roof.

Clearly the field measurement of daylight factor calls for a persevering observer, and for much patience from the occupants of the building. Even so the measured values will certainly not be accurately reproducible. One may well ask, what *is* the daylight factor inside a building? Is it the value measured *in situ*, despite inevitable departures from the C.I.E. sky luminance pattern? Is it the value measured in a scale model under an artificial sky, despite the horizon errors to be discussed below? Or is it the value calculated by, for example, the methods to be discussed in Chapters 8 and 9? If the third possibility is accepted as correct, daylight photometry is obviously an overrated occupation!

The device illustrated in Fig. 6.16 provides a radically different approach to the problem of undertaking a daylight survey [6.8]. It

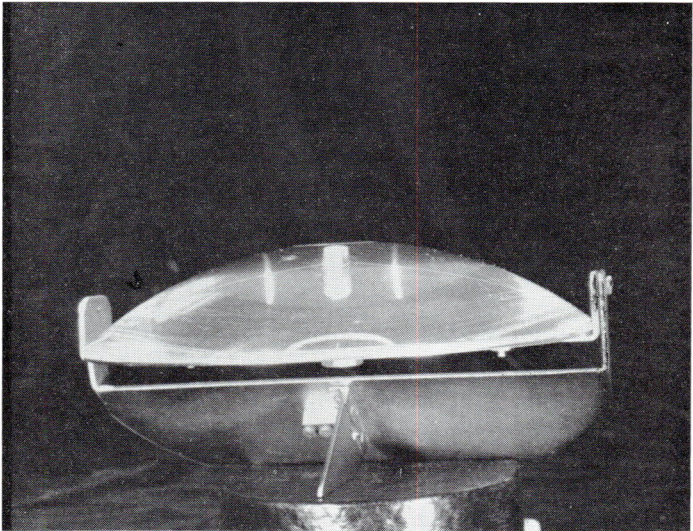

Fig. 6.16. (a)

consists of a plano-convex lens with a grid of dots pasted to its flat base. If it is placed on a table indoors and viewed from above a distorted reflection of the whole room above the level of the table top can be seen in the upper surface of the lens. The dots are so arranged that the sky factor, expressed as a percentage, is equal to

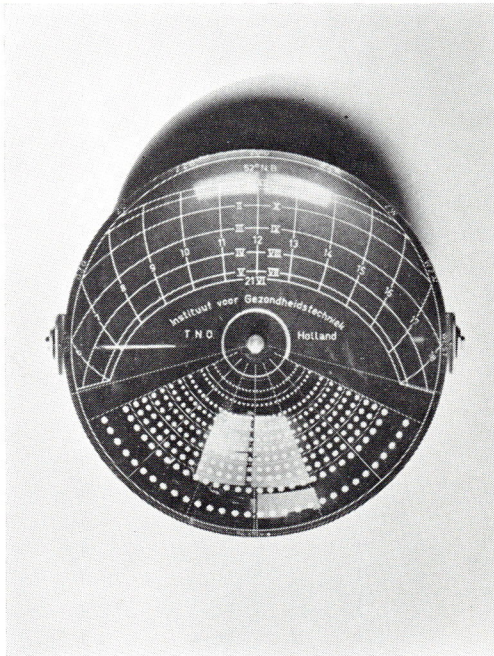

(b)

Fig. 6.16. The T.N.O. instrument (a) side view, (b) seen from above: sky factor = 7·2 per cent. The sun tracks on this instrument are for measuring insolation.

one-tenth of the number of dots seen within the reflected outline of the visible sky. The instrument gives only a rough idea of the daylight available, for it ignores the effect of glazing, dirt, interior reflections and non-uniform sky luminance, but it can be used by unskilled inspectors under any weather conditions, gives reproducible readings quickly, and can be moved around a building without disturbing the occupants.

6.4. The Measurement of Luminance

Although visual luminance meters, such as that illustrated in Fig. 6.4, are being steadily replaced by photoelectric devices they still have advantages for some forms of daylight photometry. "Disappearing spot" meters based on the simple Lummer–Brodhun cube can be designed to provide a clear view of the objects or surfaces under examination, so that the operator can locate precisely the point whose luminance is being measured. Visual meters need not incorporate portable microammeters or galvanometers and can be designed to dispense with trailing leads altogether.

The crude but robust photoelectric meter illustrated in Fig. 6.17 (a) is suitable for measuring the luminance of bright extended sources such as the sky or window, but lacks sensitivity and has a large and ill-defined acceptance angle. The instrument designed by Schreuder [6.9], whose optical system is shown in Fig. 6.17 (b), incorporates many advantages. The objective lens A focuses an image of the scene in the plane of the rotatable diaphragm C. Apertures of various shapes and sizes are cut from this diaphragm. The observer chooses an aperture to match the portion of the view whose luminance is to be measured. The flux passing through the aperture will be proportional to the average luminance of the selected area; it is measured by the photomultiplier tube D, colour-corrected by the filter K. The observer F looks through the eyepiece E and sees the whole scene reflected in the semitransparent mirrors B_1 and B_2. The selected area is also reflected by the semitransparent mirrors B_3 and B_4, and appears as a bright central area in the field of view. Since B_2 and B_4 are both semitransparent the scale and meter needle G are also visible during the measurement. The scale is illuminated by light transmitted through B_2 and reflected by the mirror H. The lighting on the scale will then be proportional to the luminance of the scene in front of the instrument. The built-in calibration unit J enables the instrument to be used with a wide range of aperture sizes. Whenever the aperture is changed the gain of the amplifier can be adjusted to restore the appropriate luminance calibration.

The advantages of Schreuder's instrument include absence of parallax when aiming, variable shape and size of measurement area, and facilities for viewing the whole scene and for reading the lumi-

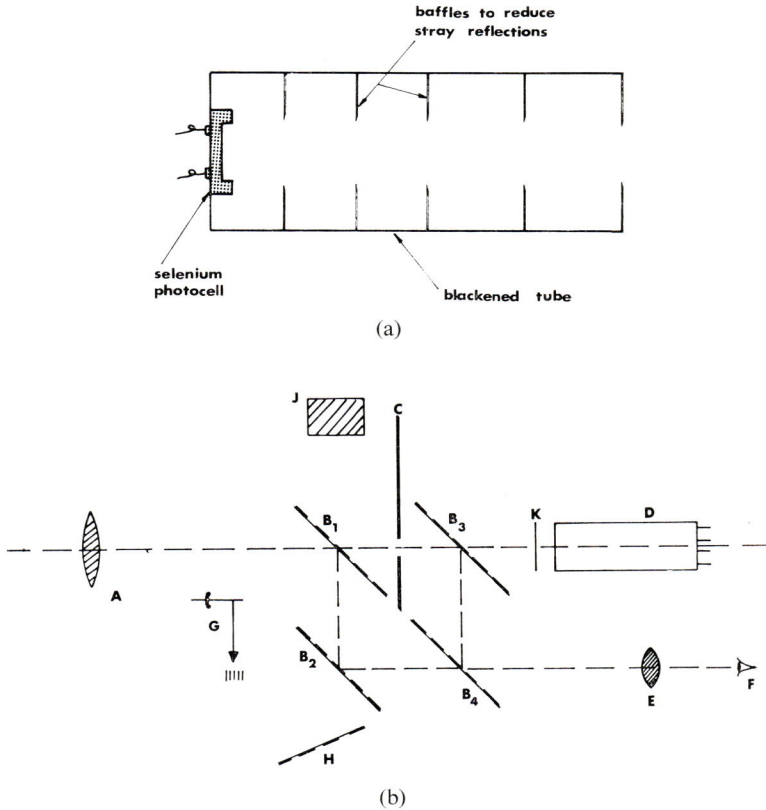

Fig. 6.17.   Photoelectric luminance meters (a) blackened tube; (b) Schreuder pattern.

nance scale simultaneously. While commercially available photo-electric luminance meters vary widely in sophistication and reliability (the worst are photographic exposure meters drafted into double service) none can hope to retain its calibration indefinitely, so the purchase of such instruments can seldom be justified unless photo-metric calibration facilities are also available.

Since a photograph is nothing more than a permanent record of a distribution of luminance it might seem convenient to measure the luminance of a scene by measuring the reflection factor of portions of a photographic print or the transmission factor of the negative under controlled conditions in the laboratory [6.10]. The accuracy will be poor since the response of photographic materials is generally non-linear and depends on the spectral distribution of the incident light. The characteristics of apparently similar films or plates vary from batch to batch, and the final result depends on how they are exposed and processed. The usual procedure is to reserve a portion of the film or plate and expose this later to an array of sources of known luminance in the laboratory; the latter serve to calibrate the whole photograph.

6.5. Photometric Calibration

All photometric measurements in a daylight laboratory will be based upon comparison with the intensity of a calibrated lamp, or with the illumination or luminance which it produces. Suitable calibrated lamps, preferably with uniplanar ("monoplane") single coil filaments, can be obtained from standardising laboratories. These are used for calibrating a hierarchy of "working standards" to be used in subsequent photometry and checked against each other at suitable intervals. They are calibrated at some specific colour temperature, generally 2854°K ("colour temperature" is a shorthand description of the spectral quality of the light emitted; it is the temperature of the Planckian radiator which matches the colour of the incandescent source in question). Lamps operated at higher colour temperatures do not retain their calibration for a useful period.

Working standards of intensity are calibrated on a photometric bench (Fig. 6.18). A note is made of the voltage at which each lamp has the same colour temperature as that of the calibrated lamps. The intensity of each lamp at this voltage is then found by applying the inverse square law. The detailed procedures, and the precautions which must be taken to ensure good screening and voltage stabilisation, are beyond the scope of this book and can be found in the standard works on photometry [6.11].

A photoelectric cell and circuit can be calibrated on the bench by placing the cell at various distances from a working standard and noting the photocurrent. The illumination at each distance is obtained by applying the inverse square law. For natural lighting, photocells should preferably be calibrated at a colour temperature of about 6500°K. This is accomplished by interposing a "colour

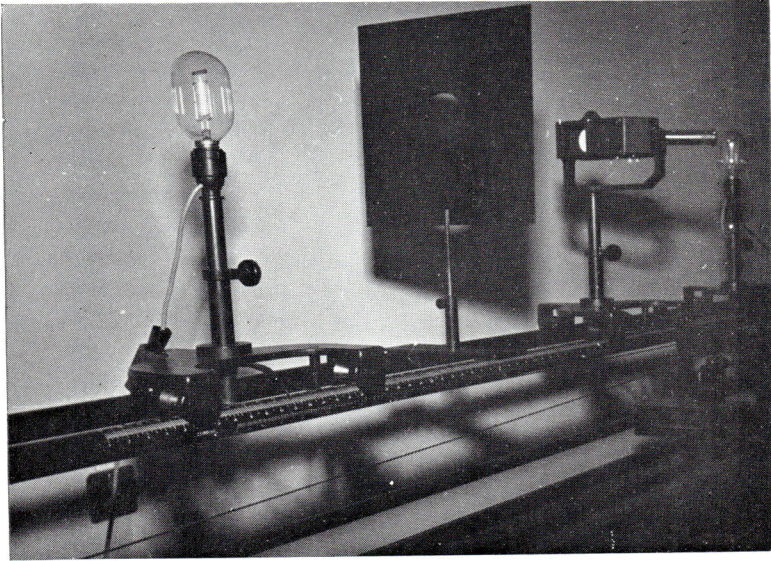

Fig. 6.18. Bench with Lummer–Brodhun photometer. Some screens have been removed to show two lamps whose intensity is being compared.

temperature conversion filter", with accurately determined transmission properties, between the lamp and the cell [6.12]. Such filters are manufactured to cover a wide range of colour temperature steps, and can also be used for checking the colour correction of a photocell.

The directional response of a scalar illumination meter is found by recording the photocurrent when the angle of incidence is varied. This method is unsuitable for checking the cosine correction of a photocell, as such a wide range of signal is involved, and departures from linearity might be of the same order as the cosine error. In this case the correct procedure would be to note the photocurrent when the cell is illuminated at normal incidence. When the cell is rotated through a known angle the lamp is moved towards the cell until the photocurrent is restored to its original value. So long as the square of the distance between lamp and cell is proportional to the cosine of the angle of incidence the cell is free from cosine error. A calibrated lamp is not, of course, needed for checking the linearity or the directional response of a cell.

A luminance meter is calibrated by measuring the luminance of a surface of known luminance factor at various distances from a lamp of known intensity. A convenient comparison surface consists of a sheet of depolished opal glass or opal acrylic plastics with an opaque black backing. Its luminance factor can be determined by comparison with a silvered surface covered with a fresh layer of smoked magnesium oxide one millimetre thick. This has a luminance factor of 1·01 when illuminated normally and viewed from an angle of 45 degrees [6.13], i.e., its luminance in foot-lamberts is 1·01 times the illumination in lumens per square foot.

6.6. Models

It was demonstrated in Section 3.1 that the distribution of illumination and luminance inside a scale model must be identical with their distribution inside the original building provided that the colours and textures of all surfaces, and the luminance of all sources, are accurately reproduced. For natural lighting the source luminance condition is easily met since the same sky illuminates both the model and the real building. Scale models have therefore been used extensively both for predicting daylight factors and for anticipating the general appearance of a daylit room. They enable alternative colour schemes or window designs to be rapidly compared.

For measuring daylight factors it is unnecessary to build an elaborate model, but thin card and balsa wood, being slightly translucent, are unsuitable for the outer walls. The scale of the model does not matter provided that the photocell is thin enough for its surface not to project above the working plane. External obstructions visible from the window can be represented by obstruction profiles subtending the same angles. If the model represents a ground floor room the ideal arrangement is to take model and photocell on to the proposed site; obstructions are then certain to be in the right place. All reflection factors should be correctly reproduced, but it is less important for hue and chroma to be accurate; indeed a neutral grey finish on the model is likely to reduce any error from poor colour correction of the photocell. The use of grey paint does not, however, obviate the need for a colour-corrected photocell, as the neutrality of window glass and of grey paint does not generally extend beyond the limits of the visible spectrum whereas the response of an uncorrected photocell may well do so.

Daylight factor readings must be made on an overcast day and are subject to the sources of error enumerated in Section 6.3, but the measurements in a model are much less inconvenient than in a real building as readings can be taken in rapid succession without the need to upset furniture and occupants. The use of architectural models for teaching, for design and for "selling" architecture to a lay client is now so widespread that it is hardly necessary to recite the advantages or the pitfalls involved in using a scale model for appraising the appearance of a daylit room. Anyone approaching this subject for the first time would be well advised to build a scale model of his own office and place the model in front of the window so that both office and model face the same outdoor scene and receive the same daylight illumination. Comparison of the model with the full-scale room will then show the effects of scale and of attention to detail in furniture, fitments and finishes. A very diminutive model has all the fascination of a doll's house and is therefore difficult to appraise objectively. An unfurnished model can give a misleading impression, as indeed can an empty full-sized room, though it may be quite suitable for measuring daylight factors. Ideally a model should be large enough for the observer to put his head inside, through an opening in the floor. His eyes must be at the correct scale height above the floor; even a full-sized room takes on a different aspect when seen from the steps of a ladder. If the observer is unable to put his head right inside the model he must use an aperture in the side, or, preferably, in a corner of the model, large enough to permit binocular viewing and to ensure that both eyes are fully adapted to the lighting inside the model. It is as hard to judge the lighting of a model by squinting through a small aperture as to appraise a real room by peeping through the key-hole.

Despite all precautions it is impossible for a model to convey with perfect accuracy the subjective effect of a full-scale interior. Our perception of space and distance depends on a wide range of monocular and binocular cues and only under unnatural conditions, such as viewing through a pinhole or through an inverted periscope, can a model be made to look like a real building. The stereoscopic effect in a model can be made to seem less unnatural by closing one eye and relaxing the accommodation of the other by viewing through a convex lens of about 2 diopters. Judgements of visual comfort under conditions of restricted vision must, however, always be open to some doubt, and it may be wiser to accept the fact that a model built for lighting studies will always look like a model, and that conclusions drawn from subjective research on models should be subject to spot-checks in full-size rooms.

6.7. Artificial Skies

The usefulness of models for predicting daylight factors is offset by the difficulty of using the outdoor sky as a light source. One may wait some time for an overcast day, and even then the sky illumination will fluctuate from moment to moment. An artificial sky, consisting of an enclosure lit by electric lamps, avoids this difficulty neatly. The "sky" luminance pattern can be kept constant, and the illumination can be varied by dimming the lamps. It should not be imagined however that an artificial sky is a vital prerequisite for daylight research or teaching. It is a convenient if costly tool, but far less essential than a photometric bench and proper facilities for calibrating photocell circuits.

Many artificial skies take the form of an illuminated vault. The one illustrated in Fig. 6.19 is designed to provide a uniform "sky" or a C.I.E. standard overcast pattern of luminance at the flick of a switch. A hemispherical shape is most commonly employed, but is not essential. An oblate ellipsoid permits closer control of the luminance distribution, while a polyhedral geodesic dome, a vertical cylinder or even a rectangular enclosure can be arranged to provide sufficiently close approximations to an overcast sky distribution for many purposes.

The inside of the sky dome should have a matt white surface which does not yellow with age. The most suitable of the commercially available finishes is a polyvinyl acetate emulsion paint containing anatase titanium dioxide pigment. The reflection factor of the paint should be reduced to about 80 per cent by adding a little ivory black water colour or Indian ink. A higher reflection factor makes the C.I.E. distribution for overcast luminance difficult, if not impossible, to achieve, because the enhanced interreflections tend to

Fig. 6.19.   Broken view of artificial sky dome.

redistribute the illumination uniformly over the surface of the dome. A translucent dome illuminated by transmitted light leaves the whole floor space uncluttered—an advantage for demonstration purposes and especially for teaching since it will accommodate a number of students comfortably [6.14].

The horizontal illumination E_H beneath a whitened hemispherical sky is given approximately by the equation

$$E_H = \frac{\rho F}{A(2 - \rho)} \tag{6.1}$$

where ρ = reflection factor of paint,

    F = flux directed upwards by lamps,

    A = floor area of sky dome.

By applying this equation one can design an artificial sky to provide several hundred lumens per square foot, though the electric power dissipated would require special provision for ventilation, especially in warm weather. However, the illumination out-of-doors, even on an overcast day, may be of the order of 2000 lumens per square foot, and this value has never been achieved under a vaulted artificial sky. This is no drawback when the sky is used for measuring daylight factors, since only *ratios* of illumination are involved. But it is obvious that a vaulted sky will never be bright enough for the subjective appraisal of glare, gloom and visual comfort in daylit rooms.

Model rooms should stand under the artificial sky at such a height that the window head is level with the "horizon" of the sky. This stops the sky from shining straight on to the ceiling of the model and producing unnatural lighting effects. The working plane will then be below the horizon and will receive slightly too little illumination under unobstructed conditions. This "horizon" error can be minimised by making the radius of the sky dome as large as practicable.

An optical remedy for this horizon error is incorporated in the "mirror sky" illustrated in Fig. 6.20. Pioneered at the Building Research Station in England, this artificial sky consists of a box whose walls are lined with flat mirrors which, by an infinite train of repeated reflections, ensure that the image of the horizon is always at the eye-level of an observer inside the box but at an infinite distance from him. The lid of the box is an opal acrylic sheet illuminated by a close array of fluorescent lamps the spacing of which can be arranged to provide an illumination comparable with that from a real sky out-of-doors, and a luminance distribution closely approximating to the C.I.E. standard overcast sky [6.15].

The high illumination available from a mirror sky makes it the best choice for research into subjective aspects of natural lighting. It is not, however, altogether free from horizon errors, for the model will "see" an infinite number of reflections of itself and of any surrounding obstructions.

The expedient of using mirrors at horizon level could, of course, be applied equally to the dome sky. Here one would use silvered reeded glass to disperse the reflected image of the model. It is doubtful, however, whether the horizon error would generally be sufficiently serious to warrant the expense this would involve.

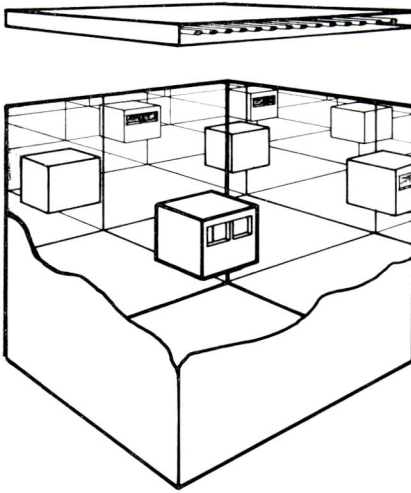

Fig. 6.20.   Broken view of box-type sky with mirrored walls.

No single artificial sky is ideally suited to every possible application. Unless requirements are clearly defined it is impossible to say which type is best for research purposes. Often more can be learned by studying models outside under a variety of weather conditions than under controlled conditions in an indoor artificial sky. If funds are limited, a greenhouse on the roof or in an open field may well give better service than a make-shift artificial substitute sky.

Chapter 7

Glazing Materials

7.1. *Transmission and the Spectrum*

The transmission factor of a glazing material will depend upon the angle of incidence and the wavelength of the light which reaches it. In Fig. 7.1 the transmission factor T of three typical transparent glasses are plotted as a function of the wavelength λ of incident radiation striking the glass normally. For a perfect neutral glass the spectral transmission curve would be a horizontal straight line. Although the sheet glass is far from neutral it will appear colourless on casual observation. The heat-absorbing glass, designed to reduce the passage of infrared radiation, has a diminished transmission factor for red and violet wavelengths; this would give the glass a bluish-green tint. Thanks to the adaptation properties of the human eye an occupant of a room completely glazed with such a glass will be unaware of the tint unless he can see the outside world through an opened window.

The curves in Fig. 7.1 are independent of the wavelength composition of the light source itself. The normal transmission factor T for a given light source will depend upon the balance of radiation among the various wavelengths within the spectrum of the light source. In general

$$T = \frac{\text{lumens transmitted}}{\text{lumens received}} = \frac{\int E(\lambda)V(\lambda)T(\lambda)d\lambda}{\int E(\lambda)V(\lambda)d\lambda} \qquad (7.1)$$

where $E(\lambda)$ = radiant power at wavelength λ,

$V(\lambda)$ = C.I.E. relative luminous efficiency at wavelength λ,

$T(\lambda)$ = transmission factor at wavelength λ.

The calculation can best be done by dividing the visible spectrum of the source into ten wavelength bands each of which contains one-tenth of the incident light flux. The average transmission factor for

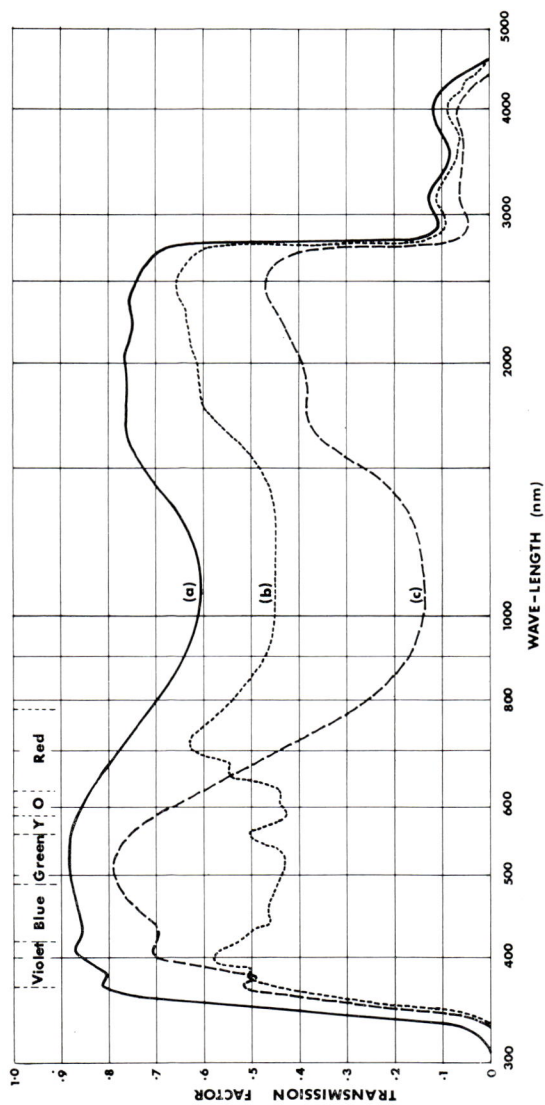

Fig. 7.1. Spectral transmission curves (a) sheet glass; (b) "grey" glass; (c) tinted heat-absorbing glass.

each band can be taken as the transmission factor for the mid wavelength of that band. The normal transmission factor T is the average for the ten bands. Table 7.1 shows mid-band wavelengths for the daylight sources whose spectra were plotted in Fig. 5.2.

Table 7.1

Mid-band Wavelengths for Daylight Sources

Direct solar radiation	C.I.E. Illuminant D_{6500}	C.I.E. Illuminant C
489 nm	485 nm	489 nm
513	511	515
526	524	530
538	535	541
548	546	552
560	556	562
570	567	572
583	579	585
599	596	601
624	622	627

7.2. Transparent Glazing

When an unpolarised ray of light containing a flux of F lumens strikes a transparent surface of refractive index n at an angle of incidence i, the proportion ρ reflected is given by Fresnel's expression:

$$\rho = \tfrac{1}{2} \left\{ \frac{\sin^2 (i - r)}{\sin^2 (i + r)} + \frac{\tan^2 (i - r)}{\tan^2 (i + r)} \right\} \tag{7.2}$$

where $n = \sin i / \sin r$.

The light entering the transparent medium will be $(1 - \rho)F$. Some of this light will be absorbed within the medium.

Figure 7.2 (a) shows a beam of light having a flux ϕ at a distance x from its point of entry into the medium. The flux $d\phi$ absorbed over the distance dx will be proportional to ϕ

$$d\phi = -b\phi dx.$$

By integration $\phi = \phi_0 \varepsilon^{-Kx}$ $\tag{7.3}$

where ϕ_0 = flux entering medium,

K = the extinction coefficient.

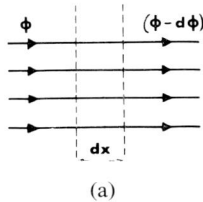

(a)

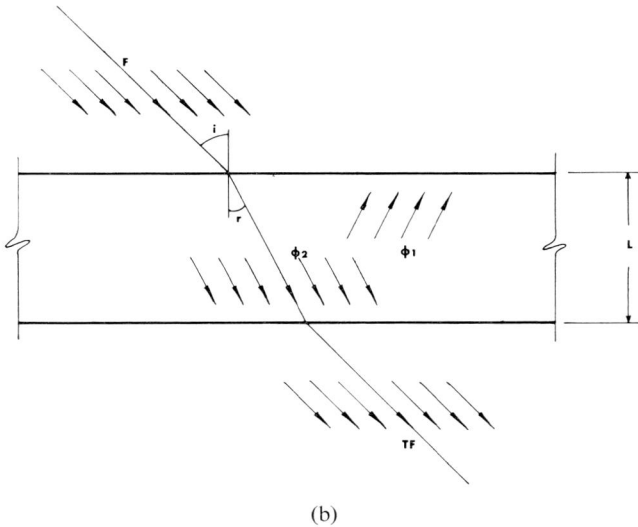

(b)

Fig. 7.2. The passage of light through a transparent medium (a) absorption; (b) interreflection.

The *internal transmission factor* T_i is defined as the ratio of the luminous flux reaching the exit surface to the flux leaving the entry surface of the medium.

In the case illustrated in Fig. 7.2 (b) the light path length from entry to exit is $L \sec r$

$$T_i = \varepsilon^{-KL \sec r} = \varepsilon^{-KL\left\{1-\frac{\sin^2 i}{n^2}\right\}^{-\frac{1}{2}}}. \tag{7.4}$$

The proportion of light reaching the exit surface which will be internally reflected is obtained from eqn. (7.2). Clearly the value of ρ is not altered when i and r are transposed.

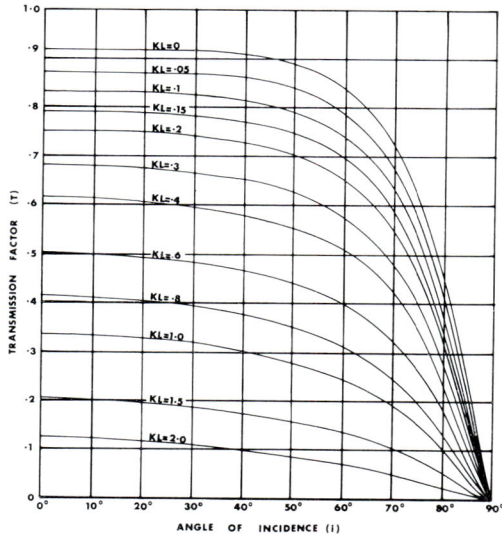

Fig. 7.3. Directional transmission curves.

An infinite train of multiple reflections now takes place between the boundaries of the transparent medium. The total flux ϕ_2 striking the exit surface is equal to

$$\phi_2 = FT_i \, (1 - \rho) + \phi_1 T_i \rho$$

where ϕ_1 = flux striking entry surface internally,

$$\phi_1 = T_i \rho \phi_2,$$

$$\phi_2 = \frac{FT_i \, (1 - \rho)}{1 - T_i^2 \rho^2}.$$

Total flux emerging = $\phi_2 \, (1 - \rho)$.

Transmission factor T of transparent medium =

$$\frac{\text{flux emerging from bottom surface}}{\text{flux striking top surface}}.$$

$$T = \frac{\phi_2 (1 - \rho)}{F} = \frac{T_i (1 - \rho)^2}{1 - T_i^2 \rho^2}. \qquad (7.5)$$

Equations (7.2) and (7.4) show that both the surface reflection factor ρ and the internal transmission factor T_i depend upon the angle of incidence i. The transmission factor T of a sheet of glass, calculated from eqn. (7.5), must therefore vary with the angle of incidence. In Fig. 7.3 the transmission factor is plotted, as a function of the angle of incidence, for glass having a refractive index of 1·52 (typical for window glass) and various values of KL in eqn. (7.4). Strictly speaking both K and n will vary with wavelength. In practice the curves in Fig. 7.3 can be applied without misgivings to any flat uncoated window glass other than purely decorative types of stained glass whose extinction coefficient can fluctuate widely over the visible spectrum.

7.3. Protective Glazing

While most windows will certainly continue to use clear sheet glass or lightly patterned glass for many years to come, various forms of protective glazing are finding increasing application.

Wired glass is often used for safety in overhead windows; although it is not mechanically stronger than sheet glass the wire mesh holds the pane together even when the glass is broken. *Laminated glass* consists of two or more sheets of glass bonded to alternate interlayers of a reinforcing material—generally a vinyl plastics material; the interlayer usually remains intact when the glass is fractured, so laminated glass is particularly suitable for shop windows containing valuables. *Toughened glass* has high compressive stresses over its surface to increase its mechanical strength and its ability to resist thermal shock; as it cannot be cut or worked after toughening any edge working or trimming must be carried out before it leaves the factory. *Rigid polyvinyl chloride* sheet has high impact strength, and, like toughened glass, does not produce jagged splinters when broken.

Various tinted glazing materials, with reduced transmission factors, have been produced to control glare by reducing the luminance of the window. *"Grey" glasses* (*see* Fig. 7.1) achieve this without seriously distorting the colours of the outdoor scene, but there is some evidence that observers prefer the colour rendering of lights containing an enhanced contribution at the deep-red end of the spectrum [7.1]. For this reason, and also because they counter the "cold" tendency of lower luminances (*see* Section 4.1), *"bronze" glasses* are becoming increasingly popular. Where permanent supplementary artificial lighting is used inside bronze glass windows the electric lamps will look relatively cold unless a warm colour of fluorescent tube is deliberately selected.

Tinted glazing materials can also be used to reduce solar heat gain by absorbing an appreciable fraction of the infrared radiation. Unfortunately the absorbed radiation heats the glass which becomes a secondary source of long-wave radiant heat. Alternative surface coatings have been devised to prevent this by reflecting, instead of absorbing, a proportion of the incident radiation while still transmitting some visible light. The reflective coating may be a film of metal or a thin dielectric coating. The latter relies upon a process of optical interference whose effect varies with the angle of incidence; the appearance of the glass therefore changes with the angle of viewing. Such reflective coatings are in their infancy and it remains to be seen whether their improved performance is sufficient to justify their extra cost when compared with heat-absorbing glass.

An intriguing possibility, still not fully developed, is a glazing material with a transmission factor which will fall automatically as the incident illumination rises. Photosensitive materials, such as photographic film, have been available for years; the difficulty consists in finding a process which is reversible, which does not deteriorate with the passage of time, and which is preferably independent of temperature, etc. One promising substance consists of a synthetic layer laminated between two sheets of glass. Heat from the sun causes the interlayer to coagulate [7.2]. The resulting non-transparent opal appearance makes this material less suitable for side windows than for roof lights, where the loss of transparency would not matter.

Double glazing offers protection from heat losses in winter and from external noise. For sound insulation the optimum spacing between the panes is about 12 inches, but for thermal insulation the optimum gap is only 0·75 inch. The second sheet of glass will, of course, reduce the transmission factor of the window.

Figure 7.4 shows a parallel beam of light containing a flux of F lumens striking the outer surface of a large double-glazed window.

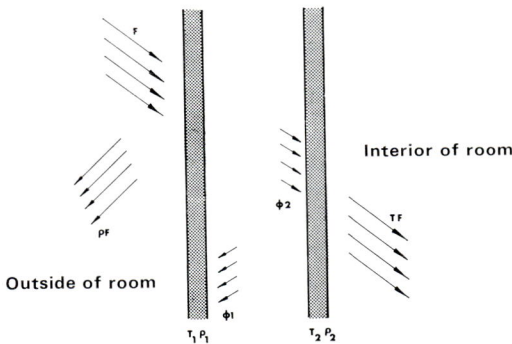

Fig. 7.4. Interreflections in double glazing.

The outer and inner panes have reflection factors ρ_1 and ρ_2 and transmission factors T_1 and T_2 respectively for the appropriate angle of incidence. The total flux ϕ_2 striking the outer surface of the inner pane is equal to

$$\phi_2 = T_1 F + \rho_1 \phi_1$$

where ϕ_1 is the total flux striking the outer surface of the inner pane as the result of multiple reflections between the two panes of glass.

$$\phi_1 = \rho_2 \phi_2$$

$$\phi_2 = T_1 F + \rho_1 \rho_2 \phi_2$$

$$\phi_2 = \frac{T_1 F}{1 - \rho_1 \rho_2}.$$

The flux emerging will be equal to $T_2\phi_2$.

Transmission factor T of double glazing $=$

$$\frac{\text{flux emerging from inner pane}}{\text{flux striking outer pane}}$$

$$T = \frac{T_2\phi_2}{F} = \frac{T_1 T_2}{1 - \rho_1\rho_2}. \tag{7.6}$$

7.4. The Diffuse Transmission Factor

The overall transmission factor of a sheet of glass due to light incident from all visible parts of an unobstructed sky will depend upon the distribution of luminance over the sky vault. The transmission factor is conveniently obtained by dividing the sky vault into ten zones whose limits are defined by their angle of incidence with the glass and chosen so that each zone contributes one-tenth of the illumination on the surface of the glass (*see* Fig. 7.5). The average transmission factor for each zone can be taken as the

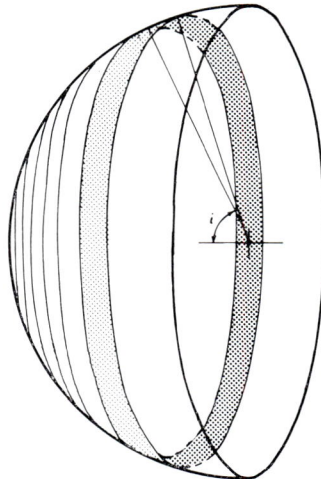

Fig. 7.5.   Each zone contributes one-tenth of illumination on glass.

transmission factor for the mid angle of the zone. The overall transmission factor is the average for the ten zones. Table 7.II shows mid-
zone angles for various glazing positions and sky luminance distributions.

Table 7.II

Mid-zone Angles of Incidence

Uniform sky	C.I.E. Standard Overcast Sky	
Glass vertical or horizontal	Horizontal glass	Vertical glass
12·9 degrees	11·4 degrees	15·9 degrees
22·8	20·2	26·6
30·0	26·7	34·05
36·25	32·4	40·3
42·15	37·8	45·95
47·85	43·15	51·4
53·75	48·75	56·9
60·0	54·9	62·75
67·2	62·3	69·4
77·1	73·1	78·3

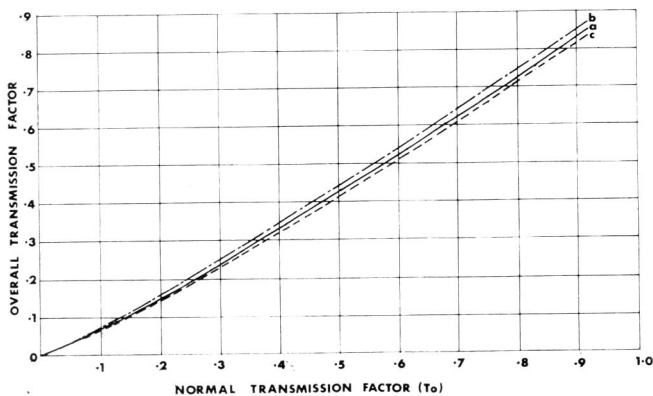

Fig. 7.6.   Overall transmission factors: (a) uniform sky; (b) C.I.E. overcast sky—
horizontal glass; (c) C.I.E. overcast sky—vertical glass.

Figure 7.6 shows overall transmission factors for each of the
sheets of glass the directional transmission factors of which were
plotted in Fig. 7.3. This enables us to estimate the overall transmission

factor of any sheet of flat glass once its normal transmission factor is known.

The diffuse transmission factor of a window is its transmission factor when it is illuminated by an unobstructed sky of uniform luminance. Table 7.II shows that mid-zone angles for a uniform sky are intermediate between those for horizontal and vertical windows illuminated by a C.I.E. standard overcast sky; the diffuse transmission factor will therefore be intermediate between the two overall transmission factors for a C.I.E. sky. Figure 7.6 confirms this result for flat window glass.

Table 7.III shows representative diffuse transmission factors for a number of colourless glazing materials. Tinted or patterned materials will have lower transmission factors.

Table 7.III

Diffuse transmission factors for various glazing materials

Transparent window glass	0·80–0·85
Clear double glazing	0·70–0·75
Patterned glasses	0·70–0·85
Wired rough-cast glass	0·80
Glass blocks	0·30–0·60
Sand-blasted glass	0·65–0·80
Acid-etched glass	0·60–0·80
Satin-finished glass	0·65–0·80
Clear acrylic plastics sheet	0·85
Clear rigid PVC sheet	0·80
Wired rigid PVC sheet	0·75
Opal acrylic sheet	0·07–0·75
Corrugated glass-fibre reinforced sheets	
Moderately diffusing	0·70–0·80
Heavily diffusing	0·65–0·70
Very heavily diffusing	0·55–0·65

Chapter 8

Roof Lights

8.1. Ceiling Dome Lights

In large factories or very deep rooms it is obviously impossible to achieve good natural lighting from side windows alone. Roof lights, on the other hand, can be arranged to provide even lighting over unlimited areas.

The most primitive roof light is merely a hole in the ceiling (Fig. 8.1 (a)), as in the Pantheon in Rome. The average daylight factor in an infinitely large room whose ceiling is perforated by a

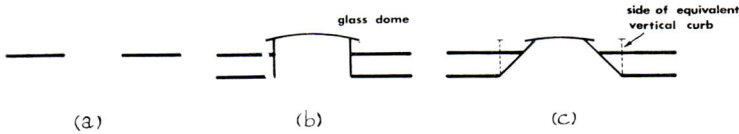

Fig. 8.1. Simple roof lights (a unglazed aperture; (b) dome with upright curb; (c) dome with splayed curb.

pattern of such holes would e equal to the area of the holes expressed as a percentage of the area of the floor. If the whole ceiling were open to the sky the daylight actor would obviously be 100 per cent.

In the more typical roof light shown in Fig. 8.1 (b) some light is lost by reflection or absor tion in the glass dome and its skirt before it can strike the wor' ng plane. The fraction of sky light striking the top of the dome hich eventually enters the factory is known as the light output rat (abbreviation LOR). This depends upon the reflection factor and oportions of the skirt, and on the transmission factor of the glazi material. Figure 8.2 shows light output ratios for rough-cast gla domes having a diffuse transmission factor of 0·84 under a C E. standard overcast sky [8.1]. Corresponding figures for other azing materials will be proportional to their transmission factors igure 8.2 reveals the importance of using a light-coloured finish o deep skirts. A light finish also reduces any discomfort that might be caused by excessive contrast between the sky-lit dome and surrounding areas of ceiling. A splayed skirt, as illustrated in Fig. 8.1 (c), will admit more light, but will shield the bright dome less effectively from direct view. The light

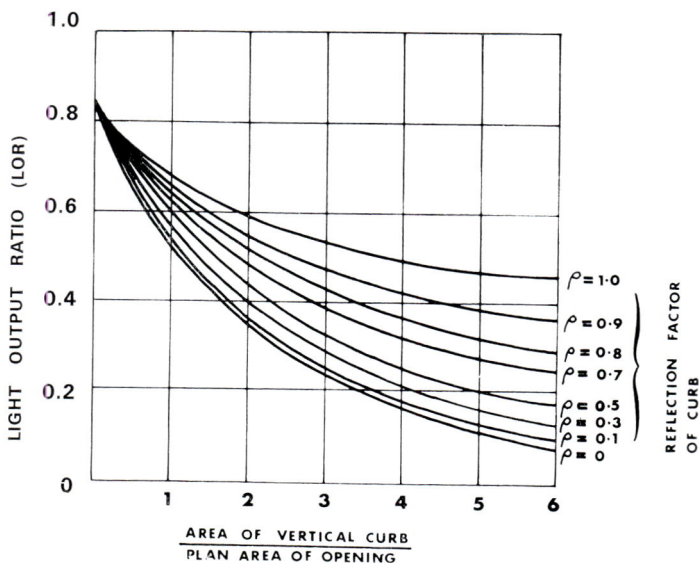

Fig. 8.2.   Light output ratio of rough-cast glass domes.

output ratio for a splayed skirt is approximately the same as for an upright curb having the same vertical height and reflection factor, but whose plan is that of the ceiling opening as shown.

The flux ϕ (lumens) entering a room through domes in the ceiling is given by the expression

$$\phi = E \times g \times \text{LOR} \qquad (8.1)$$

where E = sky illumination (C.I.E. sky) above the roof,

g = total area of glazing.

In a large room (i.e., so large that any light intercepted by the walls can be neglected) the whole of the flux included in eqn. (8.1) will reach the floor. Some of this will be reflected back to the ceiling and, as a result of the multiple reflections which ensue, the flux ϕ_F

finally striking the floor will be greater than the flux ϕ which originally entered the room (this does not break the law of conservation of energy—flux reflected from the floor may return to it again and again).

The whole of the flux reflected from the floor will strike the ceiling. We may therefore write

$$\phi_C = \rho_F \phi_F$$

where ρ_F = average reflection factor of floor,

ϕ_C = flux finally striking the ceiling.

The total flux reflected downwards by the ceiling, $\rho_C \phi_C$, will land on the floor, and the total flux ϕ_F striking the floor is obtained by adding this reflected flux to the original flux ϕ entering through the dome

$$\phi_F = \phi + \rho_C \phi_C$$

$$\phi_F = (E \times g \times \text{LOR}) + \rho_F \rho_C \phi_F$$

$$\phi_F = \frac{E \times g \times \text{LOR}}{1 - \rho_F \rho_C}.$$

The average illumination E_F on the floor is

$$E_F = \frac{\phi_F}{f}$$

where f = area of floor.

The average daylight factor is obtained by expressing E_F as a percentage of the outdoor illumination E

$$\text{Daylight factor} = \frac{E_F}{E} = \frac{\text{LOR} \times g/f \times 100}{1 - \rho_F \rho_C} \text{ per cent} \quad (8.2)$$

To find the average daylight factor in service it is necessary to apply correction factors for light intercepted by overhead obstructions and for flux absorbed by dirt on the glazing (*see* Section 8.2).

If domes are spaced too far apart the result will be general gloom punctuated by pools of daylight. We saw in Section 2.3 that, apart from reflected light, raising or lowering the light source does not affect the shape of the isolux curves. The curves expand or contract in proportion to the height of the source above the horizontal plane of measurement. Similarly, the shape of the resultant isolux curves due to two or more dome lights will depend only on the ratio of the spacing S between adjacent domes to their height H above the working plane. Thus the uniformity (ratio of minimum to maximum) of the illumination is determined by the ratio S/H, known as the *spacing/height ratio*. Recommended maximum spacing/ height ratios for various dome roof light proportions are shown in the left-hand column of Table A.8.I in the Appendix. The distance between any wall and the nearest row of domes should not exceed $\frac{1}{2}S$; if work is carried out at benches at the wall this distance should be reduced to $\frac{1}{3}S$.

The spacing/height ratio need not be interpreted too rigidly. Where the uniformity of the daylight is unimportant—in a corridor for example—the recommended ratio can safely be exceeded. But in a building where, for structural or other reasons, a much wider spacing is inevitable, some loss in flexibility must be accepted, and desks or work benches will be attracted towards the brighter areas.

In smaller rooms, whose height is not negligible compared with their length and width, allowance must also be made for light absorbed or redirected by the walls. In this case eqn. (8.2) is replaced by the more general expression.

Average daylight factor $= CU \times g/f \times M \times 100$ per cent   (8.3)

where M is the maintenance factor (*see* Section 8.2 below and Table A.8.III in the Appendix), and CU is a factor, known as the *coefficient of utilisation*, which depends on the design of the roof light, the proportions of the room, and the reflection factors of indoor surfaces [8.2].

Theoretical studies of the interreflection of light flux have dealt mainly with square rooms. If the length of a rectangular interior substantially exceeds its width, daylight factor prediction techniques for roof lights are based on an "equivalent square room", i.e., the square room whose width is such that it has the same lighting performance as the given rectangular room. This is achieved by calculating the *room index*. Interiors having a different shape, but the same room index, will, other things being equal, have approximately the same lighting performance [8.3].

For a square room, room index $= \dfrac{w}{2h}$. (8.4)

For a rectangular room, room index $= \dfrac{l \times w}{(l + w) \times h}$ (8.5)

where $w =$ width of room,

$l =$ length of room,

$h =$ heigh of centre of roof light, measured vertically from the working plane.

Table A.8.I gives tables of coefficients of utilisation for rough-cast glass dome roof lights of various proportions, with a skirt reflection factor of 0·7. For a given geometry the coefficient of utilisation of a roof light will be proportional to its light output ratio, so the effect of rectangular or splayed skirts and of different reflection factors can be found by comparing light output ratios in Fig. 8.2. Coefficients of utilisation for different glazing materials are proportional to their diffuse transmission factors.

All the coefficients of utilisation tabulated in the Appendix are based upon a typical floor of reflection factor 0·10. They can be used for other floor finishes without serious error; light reflected from the floor has little effect on the horizontal illumination since it cannot return to the working plane without suffering considerable losses on reflection from walls, roof or interior obstructions.

8.2. Factory Roofs

Table A.8.II contains coefficients of utilisation and recommended spacing/height ratios for the various roof lights commonly used in industrial buildings. The average daylight factor (d.f.) is given by the expression

$$\text{d.f.} = CU \times g/f \times M \times G \times B \times 100 \text{ per cent} \quad (8.6)$$

where CU and g/f are the coefficient of utilisation and the window/floor area ratio as before, and M, G and B are correction factors whose purpose is explained below.

While a building is in use, the daylight factor inside will fall as dirt gathers on the windows. The maintenance factor M is equal to the ratio:

$$\frac{\text{Transmission factor of window under average working conditions}}{\text{Transmission factor of clean window}}.$$

Table A.8.III shows typical maintenance factors for various situations [8.4]. These are the values of M that would normally be used in eqns. (8.3) and (8.6), but there is nothing hard-and-fast about these particular figures. It is both possible and desirable to increase the maintenance factor by cleaning the glass regularly, inside and out. Part of the architect's job is to include suitable access for maintenance from the outset.

Dirt will also have the effect of reducing the reflection factors of the walls, ceiling and floor. To allow for this deterioration one must calculate daylight factors on the assumption that reflection has fallen to the point where in practice one would clean or redecorate the room. Much will depend on the type of building and on the nature of internal finishes. Obviously more dirt would be tolerated in a foundry than in an instrument shop, and for a set outlay on cleaning one can maintain a relatively higher reflection factor on tiled or glazed walls than on surfaces with a textured finish.

Table A.8.II was prepared on the assumption that, except where all windows are vertical, single sheets of wired rough-cast glass will be used; the data for the vertically glazed saw tooth and the vertical monitor lights apply for unwired rough-cast glass. In these cases the correction factor G in eqn. (8.6) will be unity. For other forms of glazing, values of G are proportional to the corresponding diffuse transmission factors. Typical values of G are listed in Table A.8.IV.

Window frames, glazing bars, etc., will reduce the daylight factor at a point indoors roughly in proportion to their projected area as seen from that point. A sufficiently accurate allowance for this effect can be made by assuming that the correction factor B is equal to

$$\frac{\text{Actual area of glass in window}}{\text{Area of glass in aperture}}$$

Where the frame details are undecided or the net area of glass is hard to estimate, a value of 0·8 for metal windows or 0·7 for wooden windows may be taken as typical. The factor B must be reduced if necessary to take account of structural members, overhead services, cranes and other interior obstructions.

Table A.8.II shows that the simple shed roof has a higher coefficient of utilisation than other roof types. This means that, for a given daylight factor, this type of roof light needs a relatively small glazed area; winter heat loss through the glazing is therefore minimised. It is, however, harder to keep clean (*see* Table A.8.III), and also difficult to protect from sunlight, so shed roof buildings are liable to overheating in sunny weather. Recommended spacing/ height ratios for shed roof lights are shown in Fig. 8.3.

The saw-tooth roof minimises sun penetration, but additional shielding is needed, preferably by vertical louvers, if sunlight is to be completely excluded. The windows must, of course, be correctly orientated—northwards in the norther hemisphere and southwards in the southern hemisphere. Table A.8.II shows that the vertically glazed saw-tooth, in common with other roof lights having vertical windows, has a very low coefficient of utilisation for low room indices; it is therefore unsuitable for very small buildings. Vertical surfaces facing away from the saw-tooth window may be poorly lit, so fixed production lines should generally run across the lines of glazing. Where most work is done on horizontal surfaces the benches should be so arranged that light comes over the worker's left shoulder. This reduces direct glare, reflected glare and heavy shadows.

Monitor roof lights with vertical glazing provide very uniform illumination, but the low coefficients of utilisation in Table A.8.II mean that a relatively large window is needed to provide a given daylight factor. The resultant heat losses in winter offset the advantage of the small volume of enclosed air. Solar heat gain can be controlled by adjustable blinds on windows facing the sun. The flat ceiling below the monitors provides accommodation for overhead services and artificial lighting without obstructing incoming daylight. This ceiling receives no direct daylight, so light-coloured finishes should be applied to both floor and ceiling to prevent it looking gloomy.

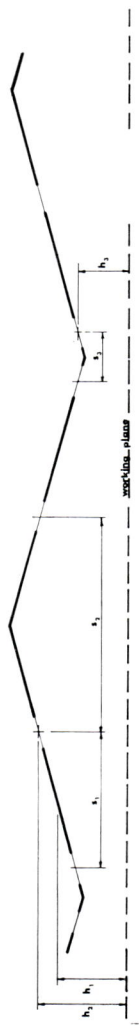

Fig. 8.3.   Maximum recommended spacing: $S_1/h_1 = 2\cdot0$; $S_2/h_2 = 2\cdot5$; $S_3/h_3 = 1\cdot0$.

8.3. Light Transfer Ratio

In Chapter 3 it was shown (eqn. (3.13)) that if AB and CD (Fig. 8.4) are sections through strips of infinite length, and AB is a uniform diffuser of luminance L (foot-lamberts), the average illumination E on CD is

$$E = \frac{AC + BD - AD - BC}{2CD} \times L \text{ (lumens per square foot)}$$

If AB is an infinitely long roof light opening on to a sky of uniform luminance L foot-lamberts, the average sky factor s.f. on the plane CD is

$$\text{s.f.} = \frac{AC + BD - AD - BC}{2CD}. \tag{8.7}$$

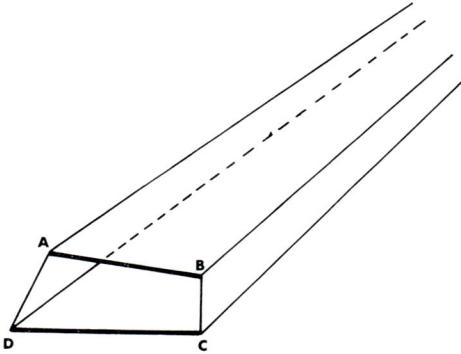

Fig. 8.4. Section through roof light of infinite length.

If a projecting ledge such as E in Fig. 8.5 hides part of AB from CD then, so long as the obstruction does not cross one of the diagonals AC and BD, eqn. (8.7) is modified to read

$$\text{s.f.} = \frac{AC + BD - BC - AE - DE}{2CD}. \tag{8.8}$$

The *light transfer ratio* (LTR) of an infinitely long roof light is defined as the average sky factor along the plane CD, divided by the ratio of the area of glazing to the area of the plane CD. It will be equal to

$$LTR = \frac{AC + BD - AD - BC}{2 \times \text{depth of glazing}}. \qquad (8.9)$$

or, in the case shown in Fig. 8.5,

$$LTR = \frac{AC + BD - BC - AE - DE}{2 \times \text{depth of glazing}}. \qquad (8.10)$$

Like the sky factor, the light transfer ratio depends only on the geometry of the window and its surroundings. It provides a rapid method of comparing the coefficients of utilisation for two long roof

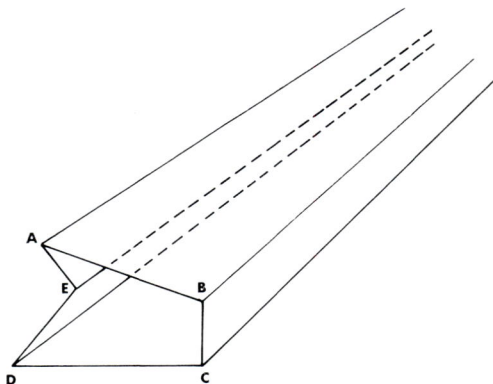

Fig. 8.5.   Roof light with projecting ledge.

lights which have the same glazing material and internal finish but which differ slightly in geometry, because their coefficients will be roughly proportional to their light transfer ratios [8.6]. Thus in Figs. 8.6 and 8.7 a uniform sky can be replaced by a uniform luminous strip AB. The line CD is drawn in such a position that direct light from the window must pass through the plane CD before reaching the inside walls or floor of the factory. The light transfer ratio can then be found from eqn. (8.9) in the case of Fig. 8.6, and eqn. (8.10) in the case of Fig. 8.7. In the latter case allowance should be made for light entering the smaller window too. This technique

is not suitable for comparing windows having widely different proportions (e.g., for comparing a shed roof with a saw-tooth) as it does
not take into account the non-uniform luminance distribution of the
C.I.E. standard overcast sky.

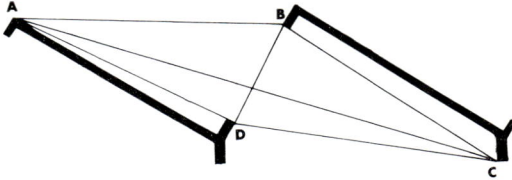

Fig. 8.6.   Light transfer ratio for saw tooth roof LTR $= \dfrac{AC + BD - AD - BC}{2 \times \text{depth of glazing}}$.

Where a roof light differs markedly from all those shown in
Table A.8.II the coefficient of utilisation for the nearest type should
be chosen. This value should then be adjusted in proportion to the
respective light transfer ratios to ensure a more accurate result.

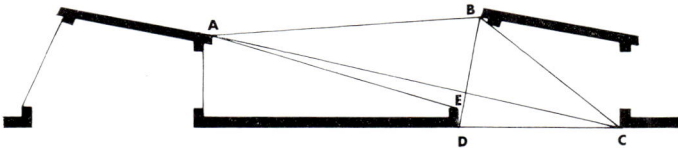

Fig. 8.7.   Light transfer ratio for sloping monitor window

$$LTR = \frac{AC + BD - BC - AE - DE}{2 \times \text{depth of glazing}}.$$

8.4. Daylight Illumination Indoors

Recommended illumination levels for a wide range of industrial tasks
can be found in the various codes of good lighting practice (*see*
Section 4.3). Once the daylight factor at a point indoors is known it is
a simple matter to estimate the sky illumination required outdoors to
ensure that the recommended illumination level is achieved inside.
Thus if the daylight factor is 5 per cent and the desired interior

illumination is 40 lumens per square foot the necessary sky illumination will be 800 lumens per square foot. In eqns. (5.4) and (5.5) sky illumination was shown to be a function of solar altitude. To estimate the times in the morning and the afternoon between which the sky illumination will be over 800 lumens per square foot one can draw a circle on the appropriate sunpath diagram (Figs. A.5.1 to A.5.7), having a radius equal to the distance of the 800 lumens per square foot graduation from the left-hand end of the radial scale for mean horizontal illumination from the sky (Fig. A.5.8). When the sun's orbit falls inside this circle the average sky illumination outdoors will exceed 800 lumens per square foot, and the indoor illumination for a 5 per cent daylight factor will exceed 40 lumens per square foot. Outside the circle the average illumination will fall below these figures.

In the United Kingdom and in most of Western Europe a hypothetical static overcast sky illumination of 500 lumens per square foot is often used as a convenient basis for daylight design. Application of the radial scale for mean sky illumination shows that at the latitude of London ($51\frac{1}{2}°$ N) this level is likely to be exceeded for about 85 per cent of the normal working year, assuming a working day of from 0900 to 1730 hours or 0830 to 1700 hours [8.7].

The "500 lumen sky" should not, however, be used indiscriminately as a basis for roof light design. An overglazed building will be difficult to heat in winter and protect from excessive sun penetration in summer; on these grounds it is seldom wise to exceed a daylight factor of 10 per cent. Localised artificial lighting can always be provided for inspection or for any other visual tasks which are too exacting to be undertaken under the available natural lighting.

8.5. Scalar Illumination from Roof Lights

The scalar daylight factor d_s at a point indoors is defined as the scalar illumination at that point, expressed as a percentage of the simultaneous horizontal illumination under an unobstructed sky of agreed luminance distribution (normally the C.I.E. standard overcast distribution). The average scalar daylight factor d_s at a given height above the floor is related to the daylight factor d.f. on a horizontal plane by the expression [8.8]

$$d_s = \text{d.f.}(K + 0\cdot5\rho'_F) \tag{8.11}$$

where K is obtained from Fig. 8.8, and ρ'_F, known as the effective reflection factor of the floor cavity [8.9], is given by

$$\rho'_F = \frac{A_F\rho_A}{A_F\rho_A + A_t(1 - \rho_A)} \tag{8.12}$$

where A_F = area of floor,

ρ_A = average reflection factor of all the room surfaces (i.e., walls and floor) below the chosen horizontal plane,

A_t = total area of all room surfaces below the chosen plane.

In most factories the area of wall beneath the chosen plane (normally eye level) is much less than the area of floor, and in this

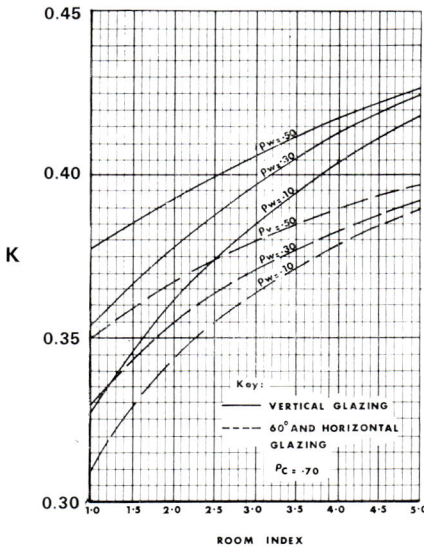

Fig. 8.8.   Values of K for calculating average scalar daylight factor from roof lights. Ceiling reflection factor = 0·70.

case the effective reflection factor of the floor cavity is simply equal to the average reflection factor of the floor.

Equation (8.11) shows the important influence of the floor cavity reflectance on the scalar illumination and hence on the general impression of brightness in a roof-lit interior. The floor cavity reflectance has a much less marked effect on the horizontal illumination, which, partly for this reason, provides a less satisfactory criterion for the subjective adequacy of interior lighting.

Experience in the United Kingdom suggests that, whatever the prevailing sky illumination, electric lights are likely to be in constant use in roof-lit buildings the average scalar daylight factor of which is less than 2 per cent. This figure therefore provides a lower limit to the useful area of glazing in a roof-lit building.

Chapter 9

Daylight from Side Windows

9.1. The Sky Component of the Daylight Factor

Figure 9.1 shows an element A of a large vertical window having a
uniform luminance of 10,000 foot-lamberts. Its intensity (candelas) in
the direction AN will be

$$\frac{10,000}{\pi} \delta y \delta z.$$

Its intensity I in the direction AO will be

$$I = \frac{10,000 \cos N\hat{A}O}{\pi} \delta y \delta z = \frac{10,000x}{\pi(x^2 + y^2 + z^2)^{\frac{1}{2}}} \delta y \delta z.$$

The illumination vector $\overrightarrow{E}$ at O will be

$$\overrightarrow{E} = \frac{I}{(AO)^2} = \frac{10,000x}{\pi(x^2 + y^2 + z^2)^{3/2}} \delta y \delta z. \tag{9.1}$$

The scalar illumination E_s will be

$$E_s = \frac{\overrightarrow{E}}{4} = \frac{10,000x}{4\pi(x^2 + y^2 + z^2)^{3/2}} \delta y \delta z. \tag{9.2}$$

The horizontal illumination E_H will be

$$E_H = \overrightarrow{E} \cos A\hat{O}Y = \frac{10,000xy}{\pi(x^2 + y^2 + z^2)^2} \delta y \delta z. \tag{9.3}$$

Figure A.9.1 in the Appendix is an elevation of the vertical
window shown in Fig. 9.1. It is drawn to a scale of eight feet to one
inch. For each square, $\delta y = \delta z = 1$ foot. The numbers in each square
are obtained by solving eqn. (9.3) for a distance x of 10 feet from the
plane of the window [9.1]. The values, which are rounded off to the
nearest whole number, show the horizontal illumination at O due to
each square foot of window.

Figure A.9.1 is handy for calculating sky factors and configuration factors but it cannot be used without modification for daylight factor calculations because it ignores two important facts; the luminance of a patch of overcast sky varies with its altitude, and the

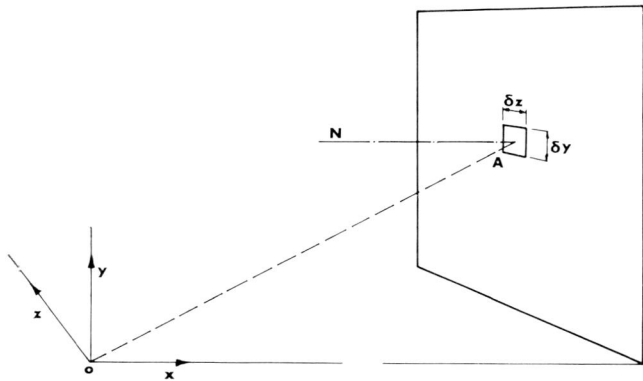

Fig. 9.1.   Sky factor at O due to element of window.

light transmitted by window glass varies with the angle of incidence. Figure A.9.2 is corrected to allow for both these factors; the sky, still providing an outdoor illumination of 10,000 lumens per square foot, now has the C.I.E. standard overcast luminance distribution, and values are adjusted for the directional transmission characteristics of typical quarter-inch flat vertical glass having a normal transmission factor of 0·88.

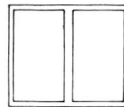

Fig. 9.2.   Window outline. Scale: eight feet to one inch.

Figure 9.2 shows the outline of a window, drawn to the same scale, eight feet to one inch. To find the horizontal illumination due to each square foot of the window at a point 10 feet back in the room one can superimpose the outline of the window on the graticule of

Fig. A.9.2, as in Fig. 9.3, making sure that y- and z-co-ordinates of the window outline are correctly placed with respect to the point O where the illumination is required. Since illumination is additive we can obtain the total illumination at O by adding together the numbers falling within the outline of the window. Where the outline cuts a square in two the number inside the square can be reduced proportionately.

The illumination found in this way is the illumination due to an overcast sky providing an illumination of 10,000 lumens per square foot on an unobstructed site in the open air—but only if the window is unobstructed, and again only if the walls and ceiling of the room are so improbably dark that they reflect no daylight to the working plane. If these conditions were fulfilled the daylight factor at the chosen point 10 feet back from the window would be equal to this indoor illumination expressed as a percentage of 10,000 lumens per square foot, or simply one-hundredth of the sum of the numbers enclosed.

To predict the daylight factor in a side-lit room we calculate separately the *sky component*, which arrives directly from the sky and the *externally reflected component*, received after reflection from outdoor surfaces such as the walls of a building opposite. The sum of these two components is then adjusted to allow for the effects of dirt on the glazing, and the multiple reflections which influence the amount and the distribution of daylight inside the room itself. Like the daylight factor the sky component and the externally reflected component are usually both expressed as a percentage of the illumination on a horizontal plane out-of-doors exposed to the whole sky.

For an unobstructed window the externally reflected component is zero. The sky component at a point 10 feet back from the window is obtained, as we saw above, by counting the numbers enclosed by the window and dividing by 100. The effect of moving the point O in the y, z plane can be found by moving the window outline with respect to Fig. A.9.2, bearing in mind that the side of each small square represents a distance of 1 foot, and that the distance x between the point O and the plane of the window is 10 feet. When O is higher than the window sill, part of the window will be below the horizon of Fig. 9.2 and can make no contribution to the sky component at O.

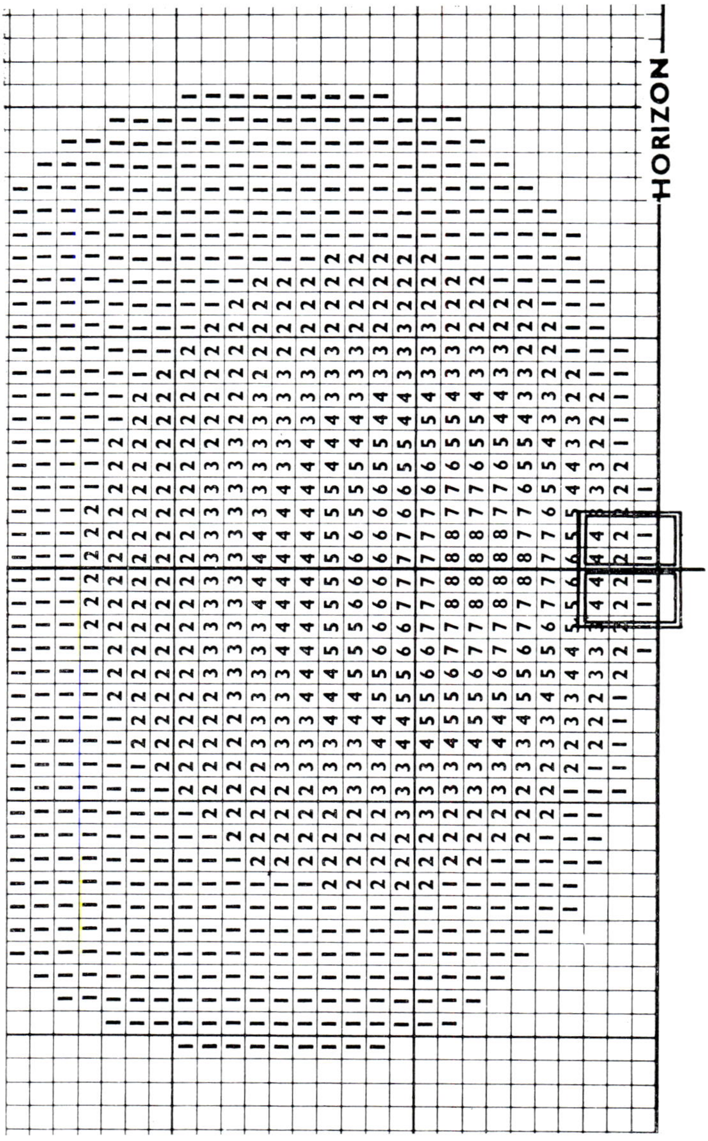

Fig. 9.3. Window outline superimposed on Fig. A.9.2. Sky component on horizontal working plane = 0·3 per cent.

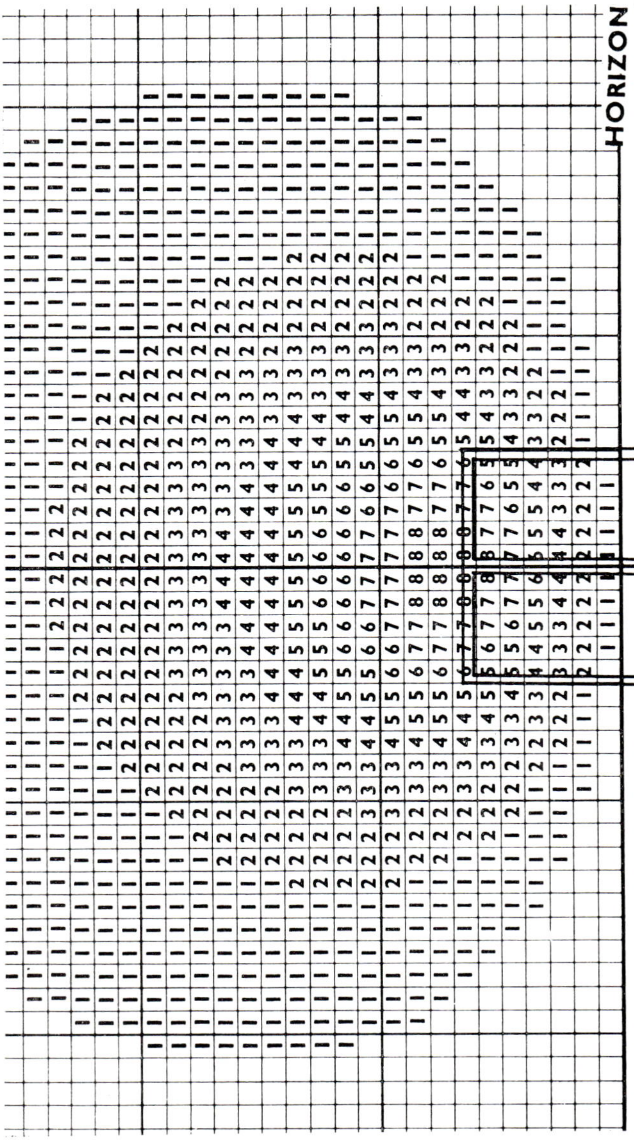

Fig. 9.4. Unobstructed window seen from a distance of 5 feet indoors. Sky component on horizontal working plane = 2 per cent.

To see the effect of changing the distance x of the point O from the window we apply the principles of perspective. Figure 9.3 can be regarded as a single-point perspective of the window, drawn with a vertical plane $1\frac{1}{4}$ inches from O and parallel to the window. Doubling x, while maintaining the same perspective distance, has the same effect, on the perspective drawing, as halving the linear dimensions (i.e., the y- and z-co-ordinates) but keeping the same perspective distance, $1\frac{1}{4}$ inches. Figures A.9.1 and A.9.2 can thus be used in conjunction with windows drawn to any scale. They are restricted not to a scale of eight feet to one inch, but to a perspective distance of $1\frac{1}{4}$ inches. To find the sky component 5 feet back from the window we redraw the window to a scale of 4 feet to one inch and use the same numbered graticule (Fig. A.9.2). This progressive halving or doubling of the window dimensions can easily be done accurately enough freehand, and, by interpolation, yields a sufficiently comprehensive picture of the sky component distribution for most architectural purposes. Occasionally one may need to know the sky component at some precise distance back from the window, say 24 feet. One could then use proportional dividers or squared paper to reduce the eight-feet-to-one-inch dimensions in the ratio 10:24.

All these procedures have assumed that the window reveals are negligibly thin. If the wall is thick, distances should be measured inwards from the *outside* surface of the wall. However, as Fig. 9.5 shows, the effective size of a window viewed obliquely is diminished by the fact that the internal reveal cuts off part of the sky. At small angles of obliquity this louvering effect can generally be ignored. On the other hand when the window is viewed very obliquely its sky component can be completely discounted, since the transmission of light through glass at grazing incidence is much diminished.

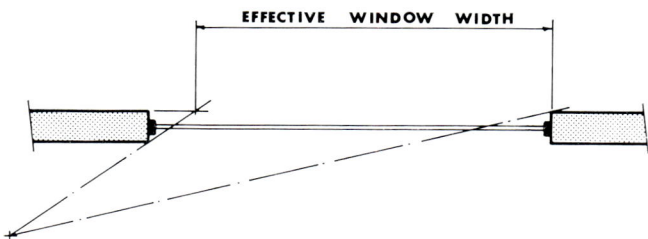

Fig. 9.5.   A thick wall can reduce the effective size of the window.

9.2. The Effect of Outdoor Obstructions

An obstruction of the type shown in Fig. 9.6 will have the same effect on the sky component as obliterating some of the squares along the bottom of Fig. A.9.2. The number obliterated depends on the height of the obstruction or, to be more exact, on the ratio of its height to

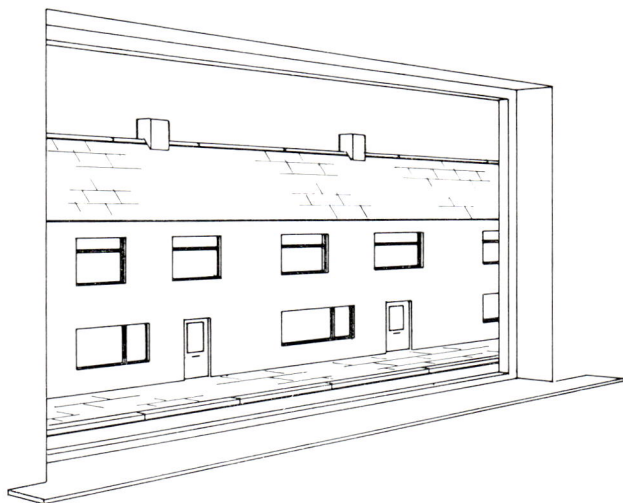

Fig. 9.6.   Long horizontal obstruction seen through window.

its distance from the point at which the daylight factor is required. Since Fig. A. 9.2 is a grid of eighth-inch squares, and refers to a perspective distance of $1\frac{1}{4}$ inches, each $\frac{1}{8}$-inch above the horizon represents a height/distance (or y/x ratio, *see* Fig. 9.1) of 0·10. Thus if the ratio $\dfrac{\text{height of obstruction}}{\text{distance of obstruction}}$ is equal to 0·40, the first four lines of squares above the horizon will be obliterated, as in Fig. 9.7.

The obliterated squares are not, however, completely lost, since the obstruction itself will reflect a certain amount of daylight into the room. This reflected contribution to the daylight factor—the externally reflected component—depends, of course, on the reflection factor of the outdoor obstructions, but, generally speaking, if they

Fig. 9.7. Sky component = 0·66 per cent. Externally reflected component = 0·12 per cent.

cover less than about a quarter of the window outline, the externally reflected component on a horizontal plane can be safely neglected. Where the outdoor obstructions are serious they can be treated as a patch of "sky" which differs from the rest of the sky only by reason of the fact that their luminance is assumed to be one-tenth of the luminance of that part of the sky which they conceal. This is strictly correct only for obstructions having a reflection factor of about 0·2 to 0·25; for lighter obstructions the externally reflected component is increased *pro rata*. To find the externally reflected component under normal conditions we find the sky component due to the hidden patch of sky, and divide it by 10.

Where the obstructions are irregular it is sometimes possible to smooth their outline to the shape of a more-or-less equivalent long horizontal obstruction. If this cannot be done it is necessary to make a perspective drawing of the obstructions outside the window, using the same perspective distance of $1\frac{1}{4}$ inches and, strictly speaking, for a viewpoint coinciding with the point at which the daylight factor is required. In practice it is seldom necessary to draw more than one or

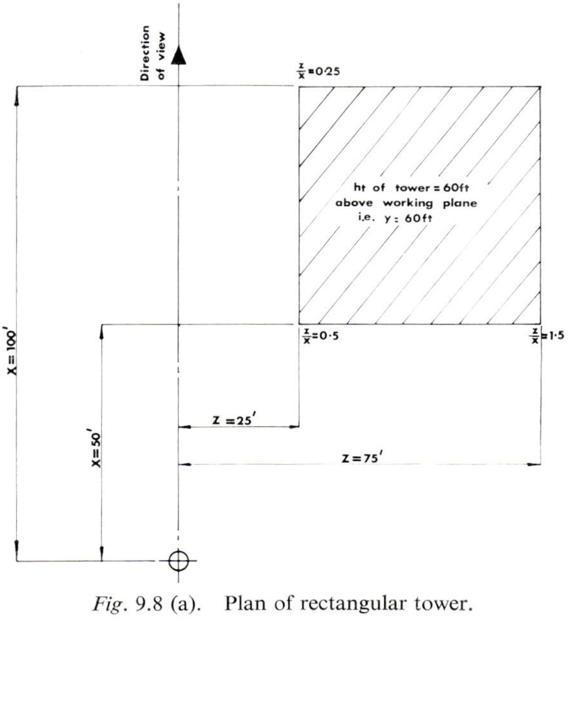

Fig. 9.8 (a).   Plan of rectangular tower.

two perspectives of outdoor obstructions; unless they are quite close to the window their outline changes very little as one moves around inside a given room.

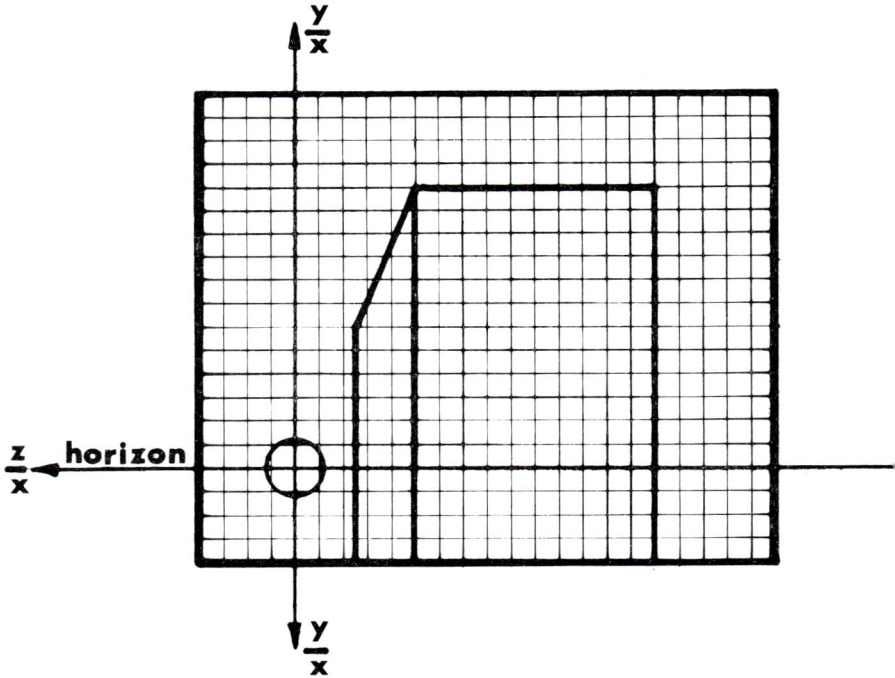

$\frac{y}{x}$

$\frac{z}{x}$ horizon

$\frac{y}{x}$

Fig. 9.8 (b).   Perspective of rectangular tower. Each $\frac{1}{8}$-in square represents a ratio y/x or z/x of 0·1.

The eighth-inch grid provides a rapid method of drawing perspectives to the required scale [9.2]. Each eighth of an inch measured upwards from the horizon represents a y/x ratio (Fig. 9.1) of 0·10, and similarly every eighth of an inch to the left or right of the origin

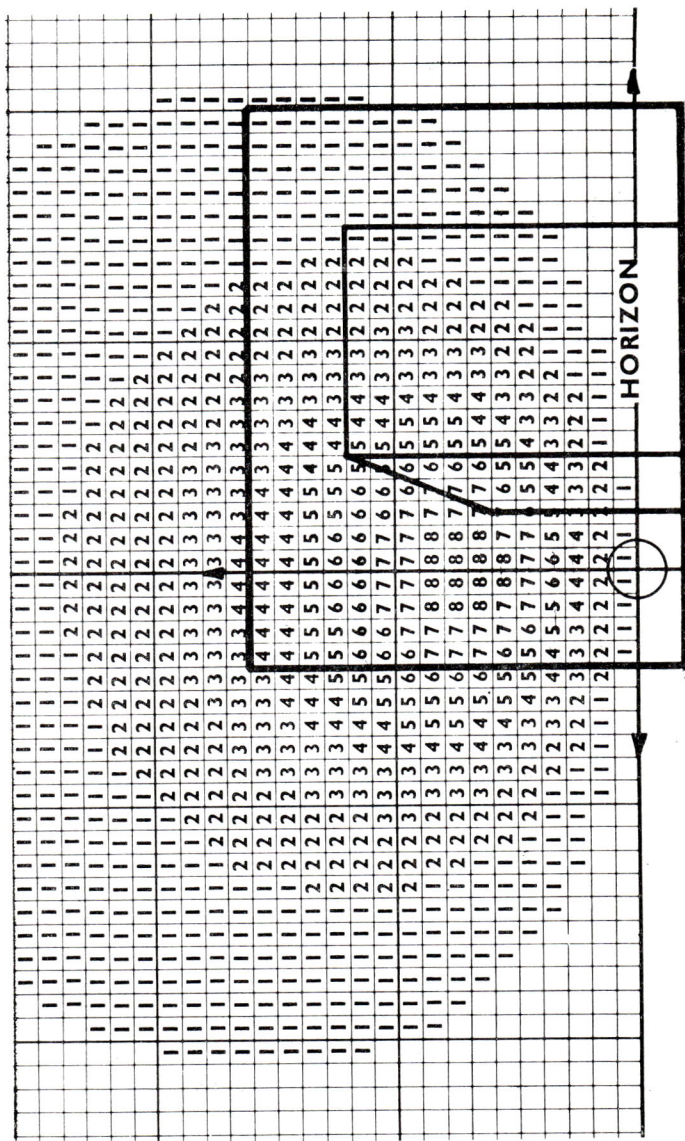

Fig. 9.8 (c). Sky component = 7·9 per cent. Externally reflected component = 0·3 per cent.

represents a z/x ratio of 0·10. Figure 9·8 illustrates a perspective drawing of a rectangular tower prepared by plotting the principal corners in this way and joining them together. To solve daylight problems there is no need to master the refinements of the art of perspective, but it is worth remembering that straight lines will be plotted as straight lines, vertical lines will remain vertical, and the shapes of all flat windows and of other surfaces parallel to the picture plane will be unchanged—they merely expand in inverse proportion to their distance from the viewpoint. For surfaces oblique to the direction of viewing, horizontal lines converge at the appropriate "vanishing point" on the horizon. Lines perpendicular to the plane of the window all converge at the centre of perspective. The outlines of complicated exterior obstructions can thus be sketched in quite rapidly, and any errors in plotting, which would distort the perspective, can be easily spotted.

9.3. The Vector Components

Equation (9.1), giving the illumination vector $\overrightarrow{E}$ due to each element of window in Fig. 9.1, enables us to predict the illumination on a plane of any given orientation at the point O. On a vertical plane parallel to the window the illumination E_x will be

$$E_x = \overrightarrow{E} \cos A\hat{O}X = \frac{10,000x^2}{\pi(x^2 + y^2 + z^2)^2} \delta y \delta z. \qquad (9.4)$$

On a vertical plane normal to the plane of the window the illumination E_y will be

$$E_y = \overrightarrow{E} \cos A\hat{O}Z = \frac{10,000xz}{\pi(x^2 + y^2 + z^2)^2} \delta y \delta z. \qquad (9.5)$$

These equations enable us to prepare grids similar to Figs. A.9.1 and A.9.2 but for estimating the illumination on vertical surfaces. These charts (Figs. A.9.3 and A.9.4) incorporate corrections for window glass transmission and for the C.I.E. standard overcast sky luminance distribution, so they can be used like Figs. A.9.2 in conjunction with the same perspective drawings of window and exterior obstructions.

One can find the magnitude and direction of the illumination vector at O by combining vectorially the three mutually perpendicular components of the illumination vector. Here it is important to remember that each of the vector components is an illumination *difference*, i.e., the difference between the illumination on the front, and the illumination on the back, of the plane concerned.

If the point O where the daylight factor is required is above the level of the window sill, the vector component of illumination E_y on the horizontal plane is obtained by subtracting light reaching the underside of the plane from light reaching the top. The "negative" component will be due to light reflected upwards from the ground outside. It is found in the same way as the externally reflected component above the horizontal plane, i.e., by assuming a definite luminance ratio, say 1:10, between the ground and the mirror image of the overcast sky. One can estimate its value by inverting Fig. A.9.2 and dividing the sum of the enclosed figures by 1000.

The vector component of illumination E_x on a vertical surface parallel to the plane of the window can be found using Fig. A.9.3 in the same manner as Fig. A.9.2. In this case light from below is *added* to light from above the horizontal.

The remaining vector component E_z, on the vertical plane perpendicular to the window, is found from Fig. A.9.4. This chart works like the others, but since the left- and right-hand sides of the chart contribute to opposite sides of the vertical plane they should be computed separately, and the smaller contribution should be subtracted from the larger to find the difference.

Once the magnitude and direction of the resultant illumination vector at a point have been found, by combining the three mutually perpendicular components vectorially, one can estimate the vector component on a surface of any given inclination—a tilted drawing board for example—by using eqn. (2.10). Unless the plane of the drawing board passes through the window this vector component will be equal to the illumination (or, as in this case, the daylight factor) on the drawing board, neglecting the light reflected on to the board by the walls, ceiling and floor of the room (the effect of interior reflections is discussed in Section 9.5 below). On the other hand it is important to remember that if direct light from the window reaches the underside of the drawing board the vector component of illumi-

nation on the plane of the board will be equal to the difference
between the illumination on the front and the illumination on the
back of the board.

9.4. Scalar Illumination and the Pepper-pot Diagrams

The scalar illumination E_s produced at a given point by a small light
source is equal to one-quarter of the direct illumination vector $\overrightarrow{E}$
The scalar illumination due to an element of a vertical window can
be found by a simple modification of eqn. (9.1):

$$E_s = \frac{\overrightarrow{E}}{4} = \frac{10,000x}{4\pi(x^2 + y^2 + z^2)^{3/2}} \delta y \delta z. \qquad (9.6)$$

Figure A.9.5, for estimating scalar illumination, is prepared on
the same basis as Figs. A.9.2, A.9.3 and A.9.4; like them, it incorporates corrections for window-glass transmission and for the C.I.E.
standard overcast sky luminance distribution.

These graticules with a figure in each square enable us to estimate sky components and external reflected components to a high
degree of accuracy, indeed to a higher degree than is likely to be
needed for architectural design purposes. Quicker but less accurate
results can be obtained by dividing the window into unequal areas
such that each area contributes a sky component of 0·1 per cent [9.3].
Figures A.9.6, A.9.7, A.9.8 and A.9.9 were obtained by dividing the
window in this manner, and placing a dot at the centroid of each
area [9.4]. These *pepper-pot diagrams* are drawn to the same scale as
are Figs. A.9.2, A.9.3, A.9.4 and A.9.5, and can be used in place of
them. To find the sky component one places a tracing of the window
outline over the pepper-pot diagram so that the origin of the diagram
marks, in elevation, the point at which the daylight factor is required.
The sky component, expressed as a percentage, is equal to one-tenth
of the number of dots enclosed by the outline of the sky; the externally reflected component will generally be one-hundredth of the
number of dots falling on the perspective of the exterior obstructions
seen through the window. The distribution of daylight throughout

the room can be explored by modifying the scale of the window outline as in Fig. 9.4, the pepper-pot diagram being substituted for the numbered grid.

9.5. Interior Reflected Light

The daylight factor at a point indoors can be increased by raising the reflection factors of the walls and of the other room surfaces. The proportion of light received by internal reflections will depend not only on the reflection factors but also on the shape of the interior.

The illumination vector in a side-lit roof is little affected by interreflections, since the magnitude of the interreflected illumination is small compared with the magnitude of the illumination vector, and equal amounts of interreflected flux reaching a point from opposite directions will produce a zero vector component. The horizontal illumination at the back of a side-lit room is generally much smaller than the illumination vector so here interreflected light may provide a substantial proportion of the daylight factor on a horizontal plane. Interior reflections contribute even more strongly to the scalar illumination, for in this case illumination arriving from opposite directions is summed arithmetically. Fortunately when the daylight factor is expressed in scalar terms the contribution due to interior reflected light, known as the *internally reflected component*, will vary very little from point to point indoors. The total scalar daylight factor at a point in a side-lit room is obtained by adding this average internally reflected component to the sum of the scalar sky component and scalar external reflected component obtained from Fig. A.9.5 or Fig. A.9.9.

The scalar internally reflected component is obtained by the *Split Flux Method* [9.5]. The daylight passing through an unglazed window of area W is split into two parts for analysis (hence the name by which this method is known):

(a) The flux ϕ_{FW} which, on entering the empty room, strikes some part of the floor or wall below an imaginary horizontal plane passing through the centroid of the window.

(b) The flux ϕ_{CW} which, on entering the empty room, strikes some part of the ceiling or wall above the horizontal plane through the window centroid.

A fraction of ϕ_{FW} and of ϕ_{CW} will be absorbed immediately by the room surfaces and can make no contribution to the internally reflected component of the daylight factor. The remainder, known as the *first reflected flux*, is reflected partly on to the other room surfaces and partly back through the window.

$$\text{First reflected flux} = \phi_{FW}\,\rho_{FW} + \phi_{CW}\,\rho_{CW}$$

where ρ_{FW} = the average reflection factor of the floor and of the wall area below the horizontal plane through the window centroid, excluding the window wall since it receives no direct light from the window,

ρ_{CW} = the average reflection factor of the ceiling and of the wall area above the horizontal plane through the window centroid, again excluding the window wall.

Let E_i = final average illumination on all the room surfaces and window, due to interior reflected light only.

The total interior reflected flux absorbed by the room surfaces or lost through the window will be $AE_i(1 - \rho)$

where A = the area of all the surfaces in the empty room, i.e., walls, ceiling, floor and window.

ρ = the average reflection factor of all the room surfaces including, this time, the window and the window wall.

To comply with the law of conservation of energy the interior reflected flux finally lost by absorption or transmission must be equal to the first reflected flux, i.e.,

$$AE_i(1 - \rho) = \phi_{FW}\,\rho_{FW} + \phi_{CW}\,\rho_{CW}$$

$$E_i = \frac{\phi_{FW}\,\rho_{FW} + \phi_{CW}\,\rho_{CW}}{A(1 - \rho)}.$$

Since ϕ_{FW} and ϕ_{CW} will both be proportional to the outdoor illumination, and to the window area, W, we may write

$$\phi_{FW} = C_1 W E_{sky}$$
$$\phi_{CW} = C_2 W E_{sky}$$

where C_1 and C_2 are constants, and E_{sky} is the prevailing horizontal illumination under an unobstructed sky, assumed overcast.

If the mean luminance (foot-lamberts) of the ground outside is taken as one-tenth of the sky illumination E_{sky} (lumens per square foot) it will be seen that

$$\phi_{CW} = 0{\cdot}5 \times W \times 0{\cdot}1 \times E_{sky} = 0{\cdot}05\, W E_{sky}$$

$$E_i = \frac{W E_{sky}}{A(1 - \rho)}\,[C_1\,\rho_{FW} + 0{\cdot}05\,\rho_{CW}]. \qquad (9.7)$$

In practice the internally reflected component is adjusted to allow for transmission losses in the glazing (a diffuse transmission factor of 0·85 is commonly assumed for clear glass). It is expressed, like the sky component and the externally reflected component, as a percentage of the horizontal illumination outdoors under an unobstructed sky.

The internally reflected component (IRC) averaged over all the room surfaces is obtained by modifying eqn. (9.7) accordingly:

$$IRC = \frac{0{\cdot}85 W}{A(1 - \rho)}\,[C\rho_{FW} + 5\rho_{CW}]\text{ (per cent)} \qquad (9.8)$$

where $C = 100 C_1$, because the IRC is expressed as a percentage.

Table 9.I

Angle of obstruction from centre of window (degrees above horizontal)	C
0° (i.e., unobstructed)	39
10°	35
20°	31
30°	25
40°	20
50°	14
60°	10
70°	7
80°	5

The value of C will depend on the extent to which the window is obstructed. Typical values [9.5] are shown in Table 9.I.

The scalar value of the internally reflected component of the daylight factor, averaged over all points inside the room, will be equal to the average value over all the room surfaces given by eqn. (9.8). The same equation is sometimes used to give a rough guide to the internally reflected component of the daylight factor on a horizontal working plane. It tends, however, to exaggerate the effect of

Table 9.II

Floor reflection factor	0·3			0·1							
Ceiling reflection factor	0·7			0·7			0·5			0·3	
Wall reflection factor	0·5	0·3	0·1	0·5	0·3	0·1	0·5	0·3	0·1	0·3	0·1
Room index*	Values of a										
1·0	1·1	1·1	1·0	1·0	1·0	1·0	0·9	0·9	0·9	0·9	0·9
1·25	1·1	1·1	1·1	1·1	1·0	1·0	1·0	1·0	1·0	1·0	1·0
1·5	1·2	1·1	1·1	1·2	1·1	1·1	1·1	1·1	1·0	1·0	1·0
2·0	1·2	1·2	1·1	1·2	1·1	1·1	1·1	1·1	1·0	1·0	1·0
2·5	1·3	1·2	1·2	1·3	1·1	1·1	1·2	1·1	1·0	1·0	1·0
3·0	1·5	1·4	1·3	1·4	1·2	1·1	1·3	1·2	1·1	1·1	1·0
4·0	1·7	1·6	1·4	1·5	1·3	1·2	1·4	1·3	1·2	1·1	1·0
5·0	2·0	1·8	1·6	1·7	1·4	1·3	1·5	1·4	1·3	1·1	1·0

* *See* eqns. (8.4) and (8.5).

ρ_{FW} which, as eqn. (9.8) shows, has a marked effect on the scalar daylight factor, but much less effect than ρ_{CW} on the horizontal illumination. In addition the internally reflected component on a horizontal plane is spread less evenly across the room, and its distribution depends greatly on the shape of the interior.

It has been found that in rectangular rooms whose walls all have the same reflection factor the *shapes* of the horizontal daylight factor contours are virtually unaffected by the reflection factors of the room surfaces. These shapes can be easily found in the case where all interior surfaces are black, and the daylight factor at each point is merely the sum of the sky component and the externally reflected component.

As the interior reflection factors are raised the internally reflected component becomes more significant, and the total daylight factor at each point is given by the formula

daylight factor $= [a \times (SC + ERC)] + [v \times e \times g/f]$ per cent   (9.9)

where a is obtained from Table 9.II,

(SC + ERC) is the sum of the sky component and external reflected component.

v is obtained from Table 9.III,

e is obtained from Table 9.IV,

g/f is equal to the ratio of glazing area to floor area.

Table 9.III

Floor reflection factor	0·3			0·1						
Ceiling reflection factor	0·7			0·7			0·5			0·3	
Wall reflection factor	0·5	0·3	0·1	0·5	0·3	0·1	0·5	0·3	0·1	0·3	0·1
Room index*	Values of v										
1·0	4·0	2·9	2·1	3·5	2·2	1·6	3·1	2·0	1·3	1·7	1·0
1·25	3·9	2·6	2·0	3·1	2·0	1·6	2·7	1·8	1·3	1·6	0·9
1·5	3·8	2·3	1·8	2·7	1·8	1·4	2·5	1·6	1·1	1·3	0·8
2·0	3·5	2·2	1·7	2·5	1·7	1·4	2·1	1·4	1·0	1·1	0·8
2·5	3·2	2·0	1·6	2·3	1·6	1·3	1·8	1·3	0·9	1·0	0·6
3·0	2·7	1·7	1·3	2·1	1·4	1·1	1·6	1·1	0·9	1·0	0·6
4·0	2·5	1·6	1·1	1·8	1·3	1·0	1·3	1·0	0·8	0·9	0·5
5·0	2·1	1·3	1·0	1·6	1·1	0·9	1·0	0·9	0·6	0·8	0·4

* *See* eqns. (8.4) and (8.5).

Table 9.IV

Angle of obstruction from centre of window (degrees above horizontal)	e
0° (i.e., unobstructed)	1·0
10°	0·9
20°	0·8
30°	0·65
40°	0·5
50°	0·35
60°	0·25
70°	0·18
80°	0·13

9.6. Correction Factors for Dirt, Glazing, etc.

It was pointed out in Section 8.2 that daylight factor calculations must allow for the effects of dirt on the glazing, of the transmission properties of the glazing medium and of light intercepted by mullions, transoms, frames, etc. Correction factors M, G and B were included in eqn. (8.6) for this purpose.

Similar correction factors are applied to eqn. (9.9) to yield the daylight factor on a horizontal plane in a side-lit room:

$$\text{Daylight factor} = \{[a \times (\text{SC} + \text{ERC})] + [v \times e \times g/f]\} \times M \times G \times B \text{ per cent.} \quad (9.10)$$

They are applied in a similar way when the daylight factor is to be expressed in scalar terms:

$$\text{scalar d.f.} = (\text{Scalar SC} + \text{scalar ERC} + \text{scalar IRC}) \times M \times G \times B \text{ per cent} \quad (9.11)$$

where, as in Section 8.2,

M = maintenance factor, from Table A.8.III,

G = correction factor, from Table A.8.IV, to be applied when materials other than clear flat glass are fitted in the window,

B = correction factor for glazing bars, etc. (*see* Section 8.2).

In dwellings, and in prestige office buildings where windows are kept almost spotless, the maintenance factor M may be taken as unity, because a transparent window will start to look dirty long before any reduction in transmission factor can be measured.

9.7. Daylight Prediction for the Tropics

In tropical regions the convention of the C.I.E. standard overcast sky is wide of the mark. By contrast with the proliferation of techniques for daylight prediction in temperate climates, calculations for natural lighting in the tropics have been sadly neglected.

In hot humid climates daylight design can still be based upon cloudy conditions, but these are so variable as to be beyond hope of standardisation. It is therefore customary to design on the basis of a sky of uniform luminance [9.6]. Figure A.9.10 is a pepper-pot diagram for a uniform sky, corrected for the directional transmission properties of clear window glass.

In hot dry climates the cloudless blue sky often has a lower luminance than the sunlit ground and buildings seen through the window. The sun, not the sky vault, is the principal source of light, so here the concept of daylight factor is abandoned and a radically different prediction technique is used. Generally the indoor illumination will be higher when the sun shines on the façade containing the window, even when the latter is protected by a screening device, than when the façade is in shadow. It is therefore wise to check the indoor illumination under the latter condition.

The step-by-step procedure for finding the illumination at a given point on a horizontal working plane at a chosen time of day is as follows:

1. Draw a perspective of the view through the window using a perspective distance of $1\frac{1}{4}$ inches, as for daylight factor prediction. If the window is shaded by fixed louvers, screens or canopies these too must be shown on the perspective.

2. Choose the orthographic sunpath appropriate to the season and latitude, from Figs. A.5.1 to A.5.7. Estimate the direct illumination E from the sun on each of the surfaces seen through the window, using the radial scale, Fig. A.5.8, for vertical illumination from direct sunlight, as described in Section 5.2. Ignore surfaces which are shielded from direct sunlight. Calculate the luminance L of each surface, using the formula $L = \rho E$, where ρ is the reflection factor of the surface, and E is the illumination it receives from the sun.

3. Count the number of dots, N_D, falling on the outline of the sky shown on the perspective, using the pepper-pot diagram for a sky of *uniform* luminance, Fig. A.9.10.

4. Estimate the illumination outdoors, E_D, from the unobstructed sky vault, using the radial scale for horizontal illumination from a cloudless sky (Fig. A.5.8). The sky illumination at the chosen point indoors will be $N_D E_D/1000$. Since the cloudless sky is relatively dark its contribution often turns out to be negligibly small.

5. Count the number of dots, N_E, falling on each of the outdoor obstructions visible through the window, using the same pepper-pot

diagram, Fig. A.9.10. The illumination indoors due to each will be $N_EL/1000$, where L is the obstruction luminance found in step 2.

6. Add the illumination from the sky to the illumination reflected from the outdoor obstructions, to get E_F, the total illumination from outdoors.

7. Estimate the horizontal illumination from direct sunlight outdoors, E_G, using the appropriate radial scale, Fig. A.5.8.

8. If the window is shielded by egg-crate screens, estimate their average transmission factor T from Fig. 9.9. This transmission factor will depend on the proportions of the egg crates and on their reflection factors [9.7]. Allowance should also be made if necessary for the thickness of the louver blades.

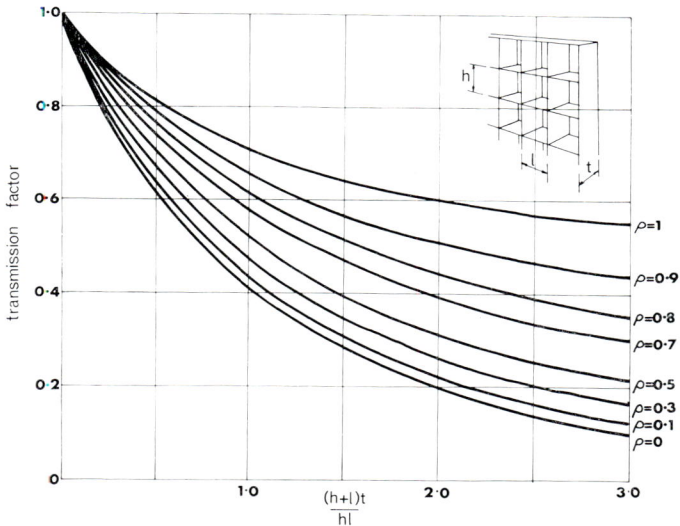

Fig. 9.9.   Diffuse transmission factor of egg-crate screen.

9. The total indoor illumination E_H at the chosen point will be

$$E_H = \{aE_F + [0.1\ vT\rho_G E_G \times g/f]\}\ M \times G \times B \quad (9.12)$$

where ρ_G is the reflection factor of the ground outside the window, and the various constants have the same meaning as in eqns. (9.9) and (9.10).

The procedure is less cumbersome than might appear at first sight, since most of the variables are constant for a given room. In practice the indoor illumination due to sunlight reflected through a well-protected window in the tropics turns out to be substantially constant throughout the day [9.8] and almost independent of orientation [9.6]. The technique is suitable for analysing a given design; unlike the methods described for an overcast sky it is quite unsuitable for designing a window from scratch. This is hardly a disadvantage under tropical conditions where visual factors must be subordinated to considerations of ventilation and thermal comfort.

Chapter 10

The Visual Environment

10.1 Daylight Criteria

The prediction techniques in Chapters 8 and 9 should enable a designer to calculate daylight factors. The problem arises: how can we specify, in these photometric terms, the lighting effects we seek to create? This is now the chief question facing daylight research workers, for many aspects of lighting, such as glitter, gloom, emphasis and visual monotony, still cannot be expressed numerically. We must beware lest an infatuation with photometric or mathematical techniques should tempt us to concentrate overmuch on daylight factors merely because we can calculate and measure them.

We saw in Chapter 4 that certain subjective effects of lighting do correlate with photometric criteria. Thus horizontal illumination is a valid index for judging the lighting of a visual task on a horizontal working plane. Scalar illumination provides a measure of the general brightness of the environment. The illumination vector indicates the dominant direction of the flow of light. The vector/scalar ratio shows the intensity of the modelling effect. Chapters 8 and 9 showed how to express each of these photometric criteria as daylight factors. What are their implications for window design?

Where roof lights can be used it is a simple matter to ensure a horizontal daylight factor adequate for most visual tasks. But where side windows only can be used, as on the lower floors of a multi-storey building, it is generally hard to get good daylighting more than a few feet inside the window without glazing virtually the whole wall from working plane to ceiling. Even when the window is unobstructed the top few feet of a window usually make the largest contribution to the sky component on a horizontal working plane, since the cosine law penalises light arriving at glancing angles. Tall thin windows are therefore better than long shallow windows of the same area, and where window size must be restricted, to reduce heat losses or traffic noise, clerestory windows will be chosen to provide a relatively high daylight factor for a given area of glass.

Many side-lit rooms are too deep to admit an adequate horizontal illumination for most visual tasks over the whole working plane. Even so, the scalar daylight factor may well be high enough to make the room look cheerfully daylit. Since the scalar illumination is unaffected by the cosine law the scalar daylight factor from a window depends mainly on the projected area of sky seen through it. The shape of an unobstructed window of given area has relatively little effect on the scalar daylight factor, so little is gained by clerestory lighting except when some visual task creates a genuine need for horizontal illumination. This is not to deny that tall windows provide pleasant lighting; their advantage lies however in the good illumination they provide in the upper part of the room, rather than in their contribution to the illumination on a horizontal working plane.

Table A.10 in the Appendix lists the minimum levels of natural lighting recommended in Great Britain by the Illuminating Engineering Society. They are expressed in terms of the daylight factor on a horizontal working plane, except where otherwise stated.

10.2. Permanent Supplementary Artificial Lighting of Interiors ("psali")

As the depth of a side-lit room increases the natural lighting looks progressively less satisfactory, for three main reasons:

1. The illumination of the window wall falls to such a point that lightness constancy breaks down when the wall surface is seen in contrast with the bright window; the resulting impression of gloom was discussed in Section 4.1.

2. The scalar illumination falls so low at the back of the room that the general effect becomes depressing.

3. The vector/scalar ratio in the back half of the room rises, producing harsh modelling.

These three effects can be remedied by using electric lighting throughout the day, especially at the back of the room. This raises the scalar illumination, reduces the vector/scalar ratio, and provides sufficient vertical illumination to restore lightness constancy to the window wall. Ideally the combination of natural and artificial lighting should be so planned as to preserve the daylit character of the interior, and this depends upon the directional effects produced, i.e., on the direction of the resultant illumination vector.

Figure 10.1 is a section through a side-lit room, showing a typical vector flow diagram. The orientation of the flow line at each point shows the direction of the illumination vector at that point.

Figure 10.2 shows the vector flow from a row of fluorescent diffusers in the ceiling. Figure 10.3 shows the effect of adding vec-

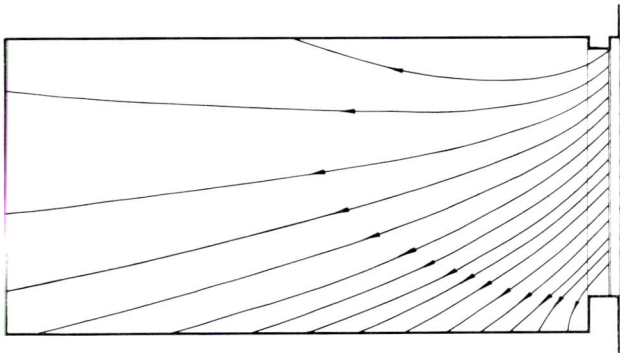

Fig. 10.1.   Flow of light through side window.

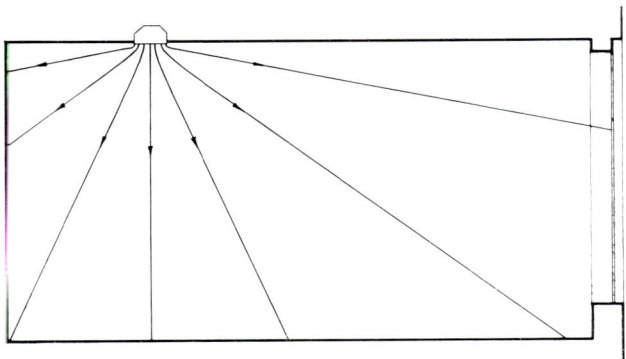

Fig. 10.2.   Flow of light from fluorescent fitting.

torially, at each point, the separate contributions of natural and artificial lighting. Although the direction of the illumination vector at the back of the room is steeper than the vector from the window alone it is still within the preferred range (*see* Fig. 4.4) and shows that the day-lit character of the interior has not been destroyed by the permanent supplementary artificial lighting. An important function of flow diagrams is to indicate the position of the neutral point, i.e., the point at which the resultant illumination vector is zero; in and around this region the modelling will be poor, so the lighting should be planned so that the neutral point is well above head height and does not coincide with significant objects inside the lighted

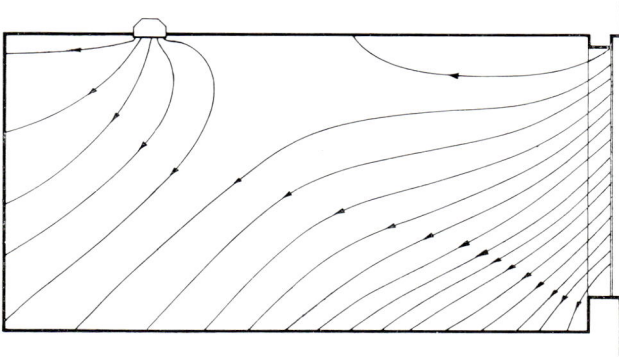

Fig. 10.3. Integrated daytime lighting.

space. Bilateral daylighting, i.e., lighting from windows in opposite walls, provides a favourable distribution of daylight factors but can give disappointing results on dull days because the neutral zone and vertically downward flow lines are likely to strike just where good modelling would be preferred.

Experiments at the Building Research Station in England have shown that permanent supplementary artificial lighting designed on the basis of an overcast sky giving an outdoor illumination of 1000 lumens per square foot will provide acceptable lighting over the whole range of weather conditions in the United Kingdom except when the outdoor illumination falls below 500 lumens per square foot [10.1]. When the sky luminance is very low the natural lighting will no longer appear dominant, and normal night-time lighting should be used in place of the special supplementary artificial lighting. To ensure that the appropriate lighting is used at all times the designer should insist that the lamps are controlled by a rotary switch labelled DAYTIME/NIGHT-TIME/OFF, or that the change from daytime to night-time lighting is operated from one central control point for the whole building.

The tendency today is to design the natural and artificial lighting together so that the same electric lighting installation can serve the dual purpose of daytime and night-time lighting. This avoids any difficulty about switching, and also dodges the debatable assumption that the designer knows better than the occupants of a room which lights should be switched on during the day.

10.3. The No-Sky Line

In very deep rooms there is so little daylight at the back of the room that it is impossible for the electric lighting to provide adequate illumination without seeming to swamp the daylight. In this case the window no longer serves any significant lighting purpose and should be designed to fulfil effectively the one visual function which remains, that of providing a link with the world outside.

Generally the bigger the window the better the view, but different portions of the view vary widely in their information content. Ideally the window head should be tall enough to ensure that all occupants of the room have an uninterrupted view of the sky-line, but once this is achieved there is little point in increasing the height of the window merely to reveal an additional patch of sky which, from the back of the room, would be counted as a source of glare.

On many urban sites the height of neighbouring buildings makes it impossible to see the whole sky-line from the back of rooms near street level. In such cases the windows should be designed to provide some view of the sky from the maximum number of indoor positions. One can draw on the plan of a room, usually at table-height, but better perhaps at eye-level, the locus of points beyond which it is impossible to see any part of the sky. This contour is known as the *no-sky line*. Where a window faces tall obstructions the no-sky line offers a rough-and-ready way of dividing a room into two parts: the area beyond the no-sky line will have an unsatisfactory view of the scene outside and will generally get insufficient daylight. For example different high-density housing layouts will generally provide different conditions of natural lighting; by plotting, on the plan of a given room, the no-sky line for each layout we can easily compare the various alternatives.

Although the sky-line is the key to a satisfactory view through a window it is by no means the only important factor. Almost invariably the scene outside is stratified horizontally into a layer of ground, a layer of city or landscape, and a layer of sky. The lowest layer, below eye level, often tells us most about activities outside. It embraces comings and goings, parks and playgrounds, traffic and pedestrians. It also contributes most to our impression of distance and thus gives scale to the surrounding townscape. Therefore although the area of window below eye level contributes little to the daylight factor, particularly on a horizontal working plane, it can be an important element in the view from the upper storeys of a tall building. Because the outdoor scene is stratified horizontally the height of the window is more critical than its width. The window proportions should be so arranged that as many occupants as possible can see some part of all three layers. They can then recognise changes of colour and texture between the layers, even where fine detail is unidentifiable [10.2]. Window width is harder to codify; the minimum width acceptable is found to be one-fifth of the diagonal of the room. If a room has only one window, and it is on one of the shorter walls, its width should be not less than one-half the width of the window wall [10.3].

Each of the various criteria discussed so far in this chapter would lead towards a different shape of window. To design a side window for a daylight factor on a horizontal working plane implies a tall window or a clerestory. To design for a scalar daylight factor implies that, in the absence of significant external obstructions, the shape of the window will matter less than its size. In both cases the lighting would be improved by room surfaces having high reflection factors. To design for an optimum flow of light, in conjunction with permanent supplementary artificial lighting, implies accepting a smaller window area in relation to the size of the room; generally the interior reflection factors have little effect on the direction of the illumination vector. To design for a view implies that the window head should not be higher than is needed to reveal the sky-line, but the sill might well be lowered to show more of the scene below eye level. Because the several criteria call for different window designs it is impossible to design a window rationally unless its purpose has been properly defined. Then the designer must reconcile the visual criteria with such other environmental requirements as the reduction of noise from traffic outside, the control of heat loss in winter and solar heat gain in summer, the control of condensation, the need for privacy, the appearance of the building elevation, and so forth.

10.4. Perspective as a Design Tool

The procedures outlined in Chapter 9, for predicting the daylight factor at a point indoors, enable us to tailor the window shape to meet whatever visual criterion is relevant. The first step is to choose

Fig. 10.4 (a)

some critical position inside the building; this could be the patient's head on a hospital bed, or the top of a desk near the back of an office. From this position draw a perspective of the outdoor scene and the

(b)

(c)

(d)

Fig. 10.4.   The view through alternative windows (a) revealing the whole sky line, (b) no sky visible, (c) some sky visible, (d) no foreground visible.

window wall using a perspective distance of $1\frac{1}{4}$ inches and a picture plane parallel to the plane of the window.

If a view through the window is the paramount criterion we can sketch a window outline to provide a satisfactory view, revealing the sky-line and also part of the scene at ground level (*see* Fig. 10.4). Alternative window shapes can be compared rapidly. By viewing the drawing through a convex lens $1\frac{1}{4}$ inches above the paper we can see all points on the perspective in the same relative directions in relation to the eye as points in the real scene would subtend when seen from the chosen point indoors. To study the effect of moving sideways from the chosen point along a line parallel to the window wall, we simply move the window outline in relation to the perspective of the outdoor scene. If exterior obstructions are very close to the building, or if the interior is very wide, it may be necessary to redraw the perspective to suit different view-points, but the gain in accuracy seldom justifies the extra work especially if the initial viewpoint was chosen wisely. To study the effect of moving towards or away from the window wall we redraw the window outline, as in Section 9.1.

Where natural illumination is the key factor in the design of the window a tracing of the perspective of the outdoor scene should be superimposed over the appropriate pepper-pot diagram, so that the centre of perspective coincides with the origin of the diagram. The window outline can then be sketched to enclose the requisite number of dots, each dot denoting a sky component of 0·1 per cent (*see* Fig. 10.5). The chart to choose will depend, again, on the purpose of the window. For the sky component on a horizontal working plane use Fig. A.9.6. For the scalar sky component use Fig. A.9.9. For modelling we need the sky component on the vertical plane facing the window—Fig. A.9.7.

Although the study of sunlight falls outside the scope of this volume it may be noted here that the same perspective drawings may be used for studying sunlight penetration. A companion volume [10.4] describes how the sun's apparent orbit across the sky may be superimposed on the perspective of the outdoor scene, as shown in Fig. 10.6. By considering the sunpaths, the sky component dots and the external view simultaneously we can devise a window to balance

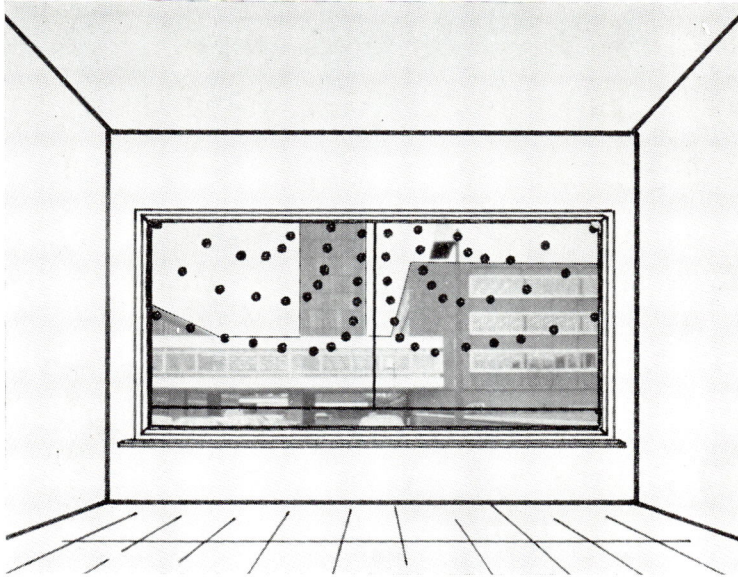

(a)

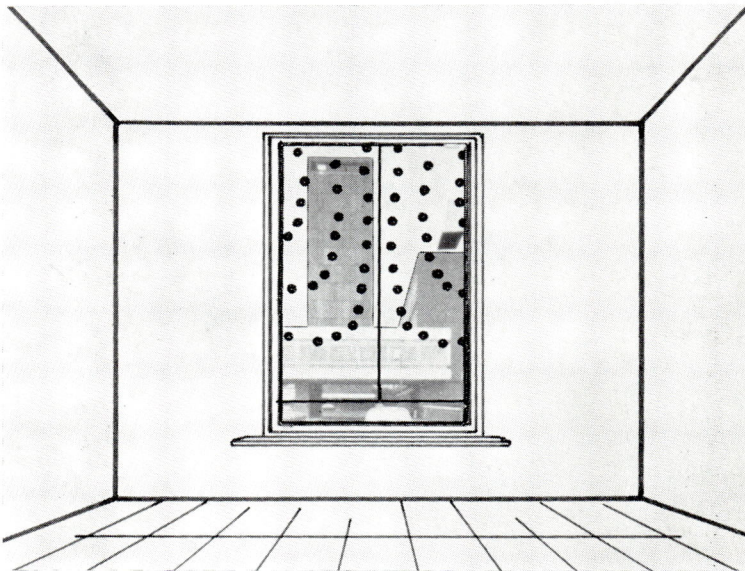

(b)

Fig. 10.5 (a)–(b). Alternative windows providing a sky component of 2 per cent on the horizontal working plane.

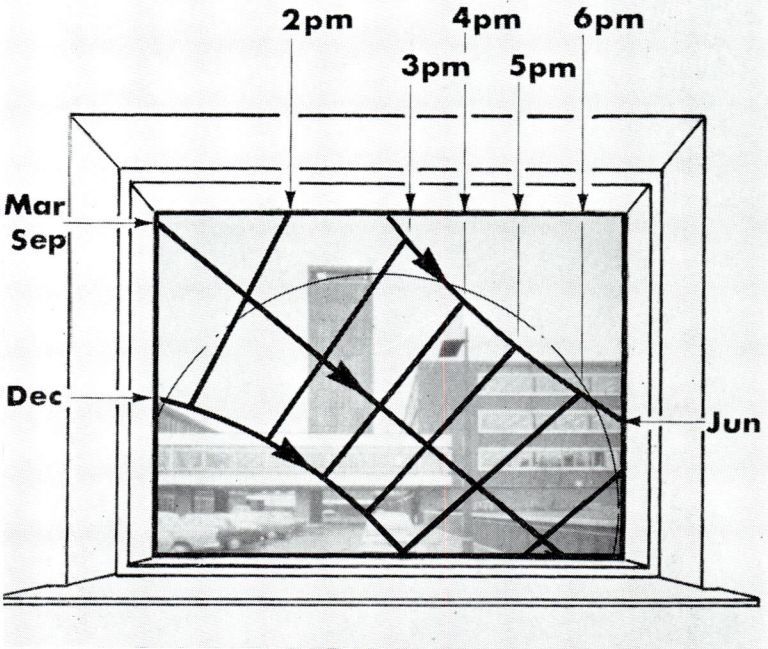

Fig. 10.6.   Solar orbits seen through a window facing 70 degrees west of south at latitude 52 degrees north. Hour lines show solar time.

the conflicting needs for sunlight, skylight and a view of the world outside.

10.5. Glare

Direct sunlight is the commonest cause of glare from windows, and the best remedy is to intercept it, by blinds, curtains, canopies, fins or louvers, before it can enter one's eyes. The sunpath ellipses, Figs. A.5.1 to A.5.7, show that in the northern hemisphere this is easy if window faces north, The sun reaches the summit of its daily t when it is due south; this means that a projecting canopy will, on the average, cast a longer shadow down a south-facing window than down a window on any other façade. During the summer such a canopy would protect the window from sunlight without serious detriment to the daylight factor; in winter the protection would be less effective, but sunlight would be less unwelcome, and some could be allowed to reach the window. East and west elevations call for adjustable shading devices which are inevitably expensive. Large windows are therefore undesirable on east and west façades. A rule-of-thumb for work places in temperate latitudes is that some complaints of glare will occur when the sun is seen through a vertical window at less than 45 degrees incidence, i.e., when it is within a $1\frac{1}{4}$-inch radius from the centre of a perspective such as Fig. 10.6.

Reflected sunlight can also be distracting, especially from a lake or river below eye level, when the eyebrows give no protection. Here a balance must be struck between comfort and prospect; this may mean raising the level of the sill.

Most of the research on glare has concentrated on trying to express its effect numerically, so that different degrees of glare—imperceptible, acceptable, uncomfortable, intolerable—can be predicted, and so that the glare produced by alternative lighting arrangements can be compared. For sunlight such refinement is superfluous. If sun shines in your eyes the glare is intolerable. But a window can cause glare even when the sun is not shining; the sky is also a source of glare, and it is useful to check the effect of sky glare especially if the window is protected from direct sunshine.

Where the layman uses the word "glare" to describe a multitude of effects typified by a bright light against a dark background, the scientist makes a point of distinguishing between discomfort glare and disability glare. Often both forms of glare are experienced together—as in the dazzle of a car's headlamps or the footlights on a stage—but it is impossible to discuss glare in quantitative terms without recognising the distinction between them, for theories and experiments dealing with one type of glare will not necessarily apply to the other.

The term "*disability glare*" describes the area of indistinct vision around a bright light. Light diffused inside the eye casts a bright veil across the scene; the brighter the veil, the more vision is impaired.

Comparisons of visual performance, with and without a glare source present, have shown that the effective luminance of this veil is roughly proportional to E/θ^2 [10.5]

where E = illumination which the glare source produces at the plane of the observer's eye,

θ = angular separation between the glare source and the object viewed.

Disability glare is "additive", i.e., the luminance of the veil produced by N identical glare sources is N times the luminance due to one of the sources acting alone. To estimate the disability glare from a large source such as a window, we treat it as a large number of small sources each with its own value of E and θ. Old people are particularly susceptible to disability glare as their ocular media are cloudier than young people's; this causes more scattering and therefore a brighter veil.

Most windows which give a view of the sky produce some disability glare. This can be observed by shielding the sky with one's hand. Details of the window frame are then seen more clearly, showing that once the glare source is hidden visual discrimination improves. This effect, which is also due to the change in the average luminance to which the eye is adapted, is seldom important, since it affects only what can be seen in the immediate neighbourhood of the window. It serves, however, as a warning against gross errors in lighting design, such as placing a chalk board right in front of a classroom window.

The term "*discomfort glare*" refers to the discomfort, as distinct from the reduction in visibility, produced by a distractingly bright source. Discomfort glare often occurs even where disability glare is absent (i.e., where no reduction in visual performance can be detected). The opposite also happens; a large window on a dull winter's day can be shown to cause disability glare when there is no obvious feeling of discomfort. The physiological basis of discomfort glare is still uncertain, but it may be associated with a saturation effect occurring when the maximum possible rate of neural response is generated in any part of the retina [10.6]. The magnitude of discomfort glare cannot be measured by the drop in visual performance, for discrimination is unaffected unless there is disability glare too. Fortunately for design purposes it is unnecessary to measure glare as such. It is sufficient to establish empirically the relative importance of the various physical factors which contribute to discomfort glare. Laboratory studies have shown that the discomfort glare caused by a bright light subtending a solid angle less than 0·05 steradian at the eye is approximately constant when $L_s{}^a\omega^b$ $f(\theta)/L_B$ is constant.

In this formula L_s is the luminance of the source, ω is the solid angle it subtends, θ is the angular distance between the source and the line of sight, and L_B is the general luminance of the background to the source. Different experimenters have obtained different values for the exponents a and b; values for a vary from 1·6 to 2·2 ,for b from 0·6 to 1·0.

The constant $L_s{}^a\omega^b$ $f(\theta)/L_B$, known as the glare constant g, will increase as discomfort increases. One can therefore set a numerical limit on the discomfort glare permissible in a given environment by specifying the appropriate glare constant. Since the exponent b is probably less than unity there is no simple way of adding together the glare effect of several sources without incurring some error. The same objection applies to treating a large source, like a window, as a number of smaller sources. However, for comparing one window with another refined data are quite unnecessary, if only because the luminance of the sky is never static. Hopkinson and Longmore estimate the glare constant for a large window facing an unobstructed overcast sky by dividing the window into a number of small elements, each equivalent to one foot square when viewed from a distance of 20 feet [10.7]. Figure A.10.1 is a pepper-pot diagram based on their results. It can be used in the same way as the other pepper-pot diagrams (see Fig. 10.7) by placing a tracing of the window over the array of dots. The constants adopted were $a = 1·6, b = 0·8$, outdoor sky illumination $= 500 \text{ lm/ft}^2$; they assumed that if the window was unobstructed the general background luminance, in foot-lamberts, would be roughly equal to the internally reflected component (IRC) of the scalar illumination obtained from eqn. (9.8). Daylight reflected by ground and buildings seen through the window will obviously mitigate sky glare by increasing the effective background luminance; where obstructions can be seen through the window the background luminance would presumably be better expressed by adding to the IRC the externally reflected component of the daylight factor on a vertical plane at the observer's eye. If N dots fall within the projected outline of a window the glare constant for an observer whose line-of-sight is normal to the window will be

$$g = \frac{10N}{ERC_{vert} + IRC}. \tag{10.1}$$

Since subjectively equivalent increments of discomfort glare are roughly proportional to logarithmic intervals of the glare constant g the British Illuminating Engineering Society formulates its glare limitations in terms of a Glare Index G, defined [10.8] as

$$G = 10 \log_{10} g. \tag{10.2}$$

Figure A.10.2 shows the Glare Index for various values of outdoor illumination, as a function of N and $(ERC_{vert} + IRC)$.

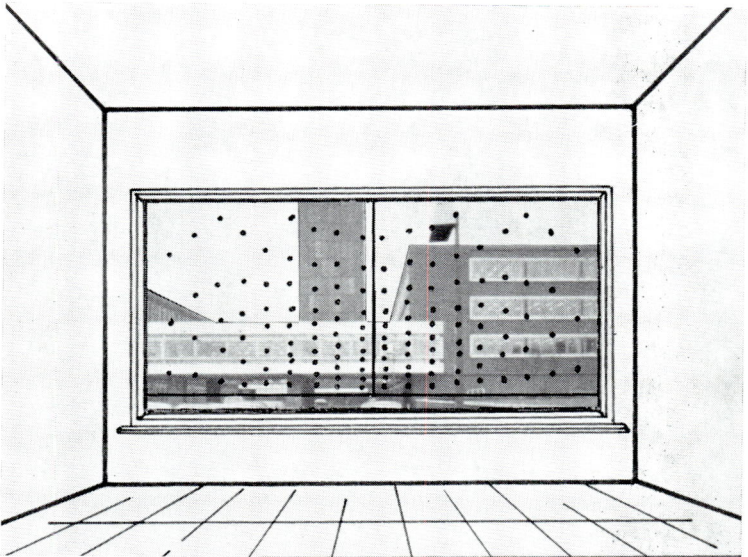

Fig. 10.7.   Discomfort glare for observer facing an unobstructed window. N = 25 dots. If $(ERC_{vert} + IRC) = 3$ per cent, Glare Index = 19 when outdoor illumination is 500 lm/ft².

The term f(θ) in the glare constant expression means that dis-comfort glare will vary as we turn our eyes. Figure A.10.3 is a pepper-pot diagram for an observer whose line-of-sight is horizontal and parallel to the plane of the window. Here the glare is diminished, and five dots in Fig. A.10.3 are equivalent to one dot in Fig. A.10.1; i.e., for this viewing condition,

$$g = \frac{2N}{\text{ERC}_{\text{vert}} + \text{IRC}}. \qquad (10.3)$$

The Glare Index is obtained from Fig. A.10.4.

Experiments on discomfort glare from artificial light sources have shown that the minimum change in Glare Index which can be identified as an increase or decrease under laboratory conditions is roughly one unit, smaller intervals being judged correctly only by chance. When the difference is 3 units observers will judge correctly nineteen times out of twenty [10.9]. Table 10.I shows the degree of

Table 10.I

Degree of discomfort glare	Range of glare index, G
Imperceptible	0–10
Perceptible	10–16
Acceptable	16–22
Uncomfortable	22–28
Intolerable	over 28

glare corresponding to various Glare Indices under laboratory con-ditions [10.8]. Different amounts of discomfort glare are obviously permissible in different places. Table 10.II shows the recommended upper limit of Glare Index for artificial lighting in a number of typical situations [10.10]. No quantitative limits have yet been set on sky glare from windows, partly because the vast majority of com-plaints about glare from windows arise from the sun rather than from the overcast sky. Experience may show that the method outlined above, albeit laborious, is sufficiently accurate to enable specific values to be codified, at least for unobstructed windows.

It would almost certainly be wrong to apply the glare limits of Table 10.II to transparent side windows, since two additional

Table 10.II

Location	Recommended limiting glare index
Most industrial tasks	25
Fire industrial tasks	22
Inspection	19
General offices	19
Drawing offices	16
School classrooms	16
School sewing rooms	13
Inspection of very small instruments	10

factors, "proximity" and "habituation", affect our experience of glare, and no experiments have yet been devised to measure their effect. The "proximity effect" arises from the fact that a large glare source just overhead feels more oppressive—even threatening—than a more distant source, e.g., the sky, much further away but having the same luminance and subtending the same solid angle [10.11]. The "habituation effect" arises from the fact that an untrained observer is much less likely to object to glare from the sky, which he has known since birth, than from an artificial source having similar luminous characteristics. Most people are quite unconscious of any discomfort glare out-of-doors unless they face the sun, but the fact that their eyes are "screwed up" shows that discomfort glare must be present, even if unnoticed.

10.6. Protection against Sky Glare

Most unshielded windows will cause some glare at some time of the year. It is therefore wise to provide curtains, blinds or shading devices of some sort before a building is occupied. Ideally they should be conceived as an integral part of the design. Permanent screens especially are too strong a visual element to be successfully accommodated as an afterthought, while movable blinds and curtains can introduce horizontal or vertical rhythms into a façade according to the architect's choice. This should not be left entirely to the whim of the client, much less to the volume of his employee's complaints.

Since most complaints about glare from windows are due to direct sunlight the best strategy for fighting glare is to protect the windows from unwanted sunlight. Shading devices outside the window have the additional merit of controlling solar heat gain. In preventing glare from the sun one automatically reduces sky glare too, often sufficiently to prevent complaints on overcast days.

Several expedients for reducing sky glare are not obvious from the glare constant expression. Splayed window reveals can buffer the sharp contrast between the bright sky and the window wall which receives no direct light from outdoors. A white-painted window wall also mitigates this contrast, especially if the wall receives additional light from windows in an adjacent wall. Neutral or tinted glasses of reduced transmission factor are less effective than might be expected since they lower the luminance of window and background in the same proportion, leaving the luminance ratio unaffected. They can however be usefully employed in the main viewing windows of a room which has clear glass in clerestories or roof lights.

Chapter 11

Daylight and Architectural Design

11.1. Urban Design

Enthusiasm for "light, space and air" was an impulse shared by almost all the pioneers of the modern movement in architecture and town planning. They condemned the traditional corridor street, lined with buildings around internal light wells, in favour of a more open layout of tall slabs or towers, which, as Fig. 11.1 suggests, should

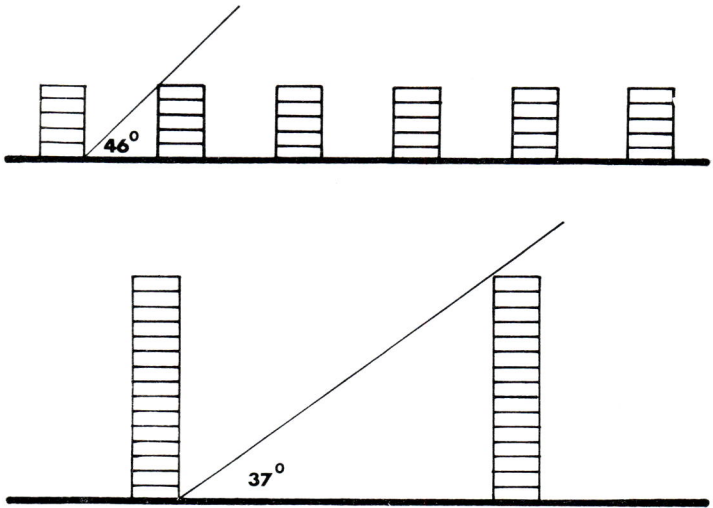

Fig. 11.1. Mutual shading by parallel buildings. Tall blocks ensure better penetration of daylight.

lead to better natural lighting. Other benefits were also expected from open planning. The space between buildings would be set free for gardens or landscaping. The spread of fire would be hindered. Upper storeys would be more remote from the noise of traffic. In Great Britain these considerations led, in 1947, to the abandonment of the customary uniform angle of setback in favour of more complex geometrical controls based on daylight requirements [11.1].

Subsequent experience has led to a measure of disillusion [11.2]. Where the tower has replaced the corridor street it has seldom released space at ground level; shops and showrooms continue to demand street frontage so the tower generally rises from a spreading podium. Even when open space is left it is often too wind-swept for comfort. Tall blocks make fire escape harder, and vertical service ducts and lift-shafts can be potential "fire chimneys". Upper storeys have not been immune from the noise of traffic, nor from the sound of aircraft. Open planning has been criticised for destroying the intimacy of the traditional urban townscape. Finally the advent of the fluorescent lamp has challenged the validity of natural lighting as a town-planning criterion. In retrospect the large-scale subordination of building layout to daylight considerations may seem ill-judged. The need remains for some alternative system to constrain building density in the interests of the physical environment.

11.2. Schools

The school is a prime example of a building type in which the post-war enthusiasm for abundant daylight has given way to caution in the light of experience. Children seated at fixed desks close to a window can suffer considerable discomfort from glare and from overheating. Here solar protection is essential, and there is a real need, as yet unfulfilled, for a tamper-proof venetian blind. Ideally a classroom should have windows in more than one side wall. Where this is impossible much can be achieved by careful detailing of the window wall. It should be finished white or off-white. Window reveals should be splayed and glazing bars tapered to reduce sharp contrasts.

In America the windowless school has been widely canvassed. In New Mexico and elsewhere a number of underground structures have been built to do double duty as elementary schools and as fallout shelters. In Michigan the windows of one school were deliberately blocked up for an experimental period to see the effect on the children and their teachers. Pupils seemed indifferent, but their teachers tended to prefer the windowless arrangement [11.3]. In

Britain the Department of Education and Science continues to require a minimum daylight factor of 2 per cent on a horizontal working plane in new school classrooms, permitting the alternative of permanent supplementary artificial lighting only under exceptional circumstances. The contrast between British and American outlooks obviously reflects climatic and economic differences, but differences in educational method and even social philosophy may be equally significant. According to Medd and Medd, British teachers "would feel very deprived if the opportunity of associating the inside and the outside were removed. Plants, animals, construction, play, earth, water, leaves, wind, rain, the sun, the moon, and passing clouds are just some of the sources of learning and experiences that can be part of the school day" [11.4].

11.3. Offices

The various alternative criteria for window design have been discussed in Chapter 10. Side windows cannot provide sufficient scalar illumination in an office to dispense with artificial lighting for most of the hours of daylight unless the office depth is severely limited. The characteristic "slab" block reflects the fact that the maximum building depth is restricted to the width of a central corridor, plus double the limited office depth.

If permanent supplementary artificial lighting is admissible, or if the windows are designed merely to provide a visual link with the world outside, a less elongated floor plan becomes feasible. A square plan makes for better site utilisation and an improved indoor environment, for most environmental problems—noise from traffic and aircraft, heat losses in winter, solar heat gain in summer—strike through the periphery of the building and can be mitigated by reducing the ratio of external wall area to floor area.

Where administrative convenience demands that dozens of people share a single office, side windows cannot hope to provide sufficient daylight for everybody and deep offices are virtually inevitable. "Bürolandschaft" layouts also call for larger office areas [11.5]. But although the fluorescent lamp has made deep buildings acceptable this does not make them universally desirable. Many employees may prefer to work in smaller offices, and supervisors may want an office of their own [11.6]. In America it is common to find daylit private offices around the periphery of a deep block, with larger windowless internal areas for less senior staff; such arrangements may prove less acceptable elsewhere. The future shape of office buildings may well depend more on these social consequences than on the daylight factor.

In board rooms and committee rooms the main axis of the table should, if possible, be perpendicular to the window wall. The chairman, at one end of the axis, should face the window. If he finds this distracting he can have the curtains or blinds, and the electric lighting, arranged to suit himself. This scheme imposes some constraint on the designer, since, for reasons of protocol, the entrance to a board room is not normally placed behind the chairman. If any other layout is adopted some of the occupants must be seen in shadow against a bright window.

11.4. Hospitals

Most hospitals in use today were built at a time when windows along opposite sides of every ward were thought essential for cross-ventilation. This guaranteed abundant natural lighting, but led to the familar "finger" plan of long wards spidering over the building site, and long journeys on foot for the medical staff. The tendency today is for wards to be daylit from one side only. This benefits the nurses and the patients too, for the latter can turn over to face away from the windows. Clearly, however, it presents lighting problems, for there is a limit to the depth of a ward which can be properly daylit, so the capacity of one ward must be limited to four or, at most, six beds. Economy in service distribution systems leads to these wards being grouped in a "race track" plan around a windowless internal service area. Hospital planning is still in a state of flux, for the compact plan does not provide for the flexibility which medical advances demand. Future trends may well be towards an "indeterminate" architecture based upon a linear plan with provision for systematic expansion [11.7].

Unilateral fenestration in the ward is probably here to stay. The window should be designed as much for its psychological effect as for its lighting function. The width of the window is likely to matter more than its height; although most patients can extend their view vertically by sitting up or lying down they have little opportunity of moving across the window so the sides of the opening set a firm limit to the field of view. In a multi-storey block more benefit may be gained from a view of the ground, where leaves and flowers are growing, than from a distant landscape; the window sill should be lowered accordingly.

Protection from direct sunlight is better provided by fixed fins or canopies than by venetian blinds. The latter may harbour dirt, they cannot be adjusted by bedridden patients, and nurses cannot be expected to judge what setting will be best for each patient, especially as the sun is in continual motion. Outdoor shading can also provide a foothold for window cleaners. If windows were cleaned from indoors dislodged dirt would blow into the ward.

11.5. Museums and Art Galleries

Sculpture is best seen in sunlight. If it *must* be exhibited indoors the lighting should aim to simulate the directional qualities of sunshine A large side window can be satisfactory. A roof light directly overhead is a very poor substitute. Clusters of spot lights, casting multiple shadows, should be avoided. One powerful light source at an altitude of 45 degrees is acceptable, with a light background to soften the shadows and provide a vector/scalar ratio in the range 1·5 to 2·0.

Jewellery and metalwork are enhanced by an abundance of daylight, but for many other museum objects there is a conflict between the demands of display and of conservation. Both visible light and, especially, ultraviolet radiation can damage textiles, paintings, leather and paper, so the amount of light striking these materials must be severely limited. Ultraviolet radiation can be eliminated from daylight by using special glass [11.8] or by fixing plastics filters inside the windows [11.9]. There is no "safe" level of illumination for sensitive materials. Any light must cause some deterioration, so whatever limitation is imposed on the illumination level must be to some extent arbitrary. In France a maximum illumination of 50 lumens per square foot is prescribed for the artificial lighting of oil paintings, and 30 lumens per square foot for pastels, water colours and other critical materials [11.10]. In the National Gallery, London, and in recent galleries in Portugal and Australia, the illumination on pictures is restricted to 15 lumens per square foot [11.11].

The law of reciprocity is generally assumed, i.e., ten units of illumination lasting for one hour will have the same effect as one unit for ten hours, so it is legitimate for daylight to exceed these limits on the brightest days. Nevertheless it is clearly easier to meet such restrictions by relying upon electric lighting, and providing occasional small windows to relieve the visual equivalent of "museum feet". Since the subtle variations of daylight have some undoubted attractions for art gallery lighting the best arrangement for a multi-storey art gallery may be roof lights on the top floor, designed to limit the entry of sunlight and to exclude ultraviolet radiation, and near windowless rooms below to house exhibits which might suffer from too much light.

Art gallery lighting should aim to emphasise the paintings. The background should therefore be slightly darker than the paintings themselves. Clerestory lighting above the paintings will be distracting. Side windows facing the paintings will be reflected into the observer's eyes. Roof lights should be carefully designed to light the pictures, rather than the floor, from such a direction as to avoid unwanted reflections [11.12].

Since illumination levels and daylight factors inside an art gallery must be kept low it is important to provide a smooth gradation of lighting from the entrance to the interior space to prevent an impression of gloom on entering the gallery.

11.6. Churches

One cannot design a window before one has defined its purpose. This was one of the themes of Chapter 10, and nowhere does it apply more plainly than in a church. For the non-liturgical type of service, daylight should be planned to reveal the face of the preacher (here the scalar illumination, the vector direction and the vector/

scalar ratio should be checked); horizontal daylight factors must everywhere be sufficient for reading hymnals. Where the emphasis is on liturgical action the natural lighting should be designed accordingly, to focus attention on the altar and the sanctuary. The illumination in other areas should be subordinated to this end, and can be "topped up" by artificial lighting when occasion demands. Light should be directed at the vertical faces of the altar. Its horizontal surface is almost hidden from the congregation, so light from directly above is largely wasted. In churches with a central altar and worshippers on all sides, care must be taken to highlight the altar without exposing the congregation to excessive glare. An alternative treatment in an axial church is to light the wall behind the altar as brightly as possible, by windows in the side walls. The altar is then revealed in dramatic silhouette, though sculptured and coloured furnishings must lose in clarity. Windows behind the altar are a source of distraction and, from a lighting point of view, quite misplaced.

Stained glass imposes architectural constraints precisely because it demands a reversal of all the normal rules for avoiding glare. Stained glass is best seen from a darkened interior. Any light from other windows reaching the surface or surrounds of the stained glass will detract from its brilliance. If the other windows have clear glazing the effect can be fatal [11.13]. If stained glass is to be used its demands must have precedence over all other lighting considerations. Single-mindedness is the key to success.

11.7. Sports Halls and Swimming Pools

Most sports are essentially outdoor activities. Indoors they are protected from bad weather, but an important function of the windows in a sports hall may well be to retain the psychological link with the open air. So far as lighting and safety are concerned, roof lights or clerestories have every advantage over side windows, but they do not provide the vital view of the world outside. Unless the surroundings are attractive it may be wise to dispense with side windows. If the view is worth retaining, side windows should be carefully designed to frame it, and trees or shrubs should be planted to mitigate glare. Large side windows are hard to clean and make a sports hall very noisy.

The most difficult problems arise in swimming pools, where the visibility of submerged swimmers may be literally a matter of life and death. Bright images of the windows, reflected by the choppy surface of the water, can hide the bottom of the pool. Windows should be so arranged as to minimise this effect. Roof lights ensure good penetration of daylight below the water surface. Side windows should, as far as possible, be situated behind the backs of the attendants. The bottom of the pool should be as light as possible. The theoretically ideal arrangement would be to install underwater floodlights, at least at the deep end of the pool. These should be a few inches below the water level, so that light reaching the surface of the water is reflected downwards. Unfortunately, cheap installations are not tamper-proof, and underwater lighting has come to be regarded as a luxury, to be used only for aqua shows or swimming pageants.

Glare from daylight is unlikely to be a source of discomfort to swimmers, as is shown by the crowds who flock to the sea on a sunny day. Great care should be taken, however, to protect divers and spectators from direct and, especially, reflected glare. The needs of spectators are best met by roof lights, or by side windows behind the spectators' heads. White tiled pool surfaces, with darker treatments elsewhere, help to focus attention on the pool.

11.8. Circulation Spaces

Side windows lend interest to a corridor through the succession of changing vistas of the outside scene. A window in the end wall of a corridor will make objects indoors appear silhouetted against a bright background, much as in a lighted street at night. Details will be hard to pick out, so this arrangement is not recommended unless the view beyond is sufficiently rewarding [11.14].

If shallow-skirted roof lights are used in a corridor the lighting will not seem unduly patchy unless the spacing/height ratio exceeds 2:1. To plan a roof light layout for a corridor, start by tentatively placing domes to mark corners, doorways and staircases. Adjust their positions so that where possible one dome does the work of two. Then add extra domes where required to avoid excessive spacing.

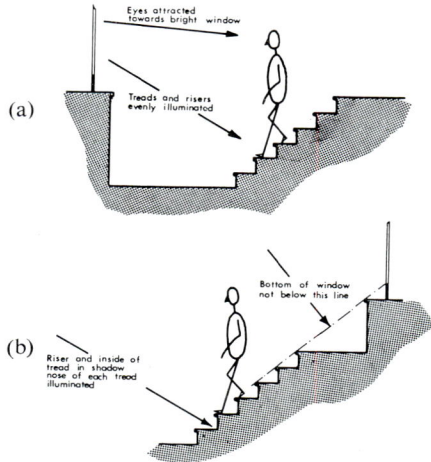

Fig.11.2.   Daylight on a staircase (a) wrong, (b) correct.

Bad lighting on a staircase can be dangerous. Fig. 11.2 (a) shows the commonest mistake. Figure 11.2 (b) shows the preferred arrangement, where each riser is in shadow, and only the nose of each step receives unobstructed light from the window.

References

Chapter 1

1.1.   P. JAY: Artificial Lighting. *Architects' Jnl.* (Jan. 4th., 1967), pp. 13–47.
1.2.   J. W. T. WALSH: *Photometry* (3rd Edition), Constable, London, 1958, Chapter 7.
1.3.   Y. LEGRAND: *Light, Colour and Vision*, Chapman & Hall, London, 1957, Chapters 4 and 6.
1.4.   H. G. SPERLING: An experimental investigation of the relationship between colour mixture and luminous efficiency. *N.P.L. Symposium on Visual Problems of Colour*, H.M. Stationery Office, London, 1958, pp. 251–277.

Chapter 2

2.1.   H. ZIJL: *Large-size perfect diffusors* (2nd Edition), Philips Technical Library, Eindhoven, 1960, Chapter 15.
2.2.   J. A. LYNES, W. BURT, G. K. JACKSON AND C. CUTTLE: The Flow of Light into Buildings. *Trans. Illum. Eng. Soc. (London)*, Vol. 31 (1966), pp. 65–91.
2.3.   J. W. T. WALSH: Modern English usage for the lighting engineer. *Light and Lighting*, Vol. 49 (1956), pp. 256–259.
2.4.   J. G. HOLMES: A method of plotting isolux curves. *Light and Lighting*, Vol. 39 (1946), pp. 158–160.
2.5.   A. A. GERSHUN: The Light Field. *Jnl. of Maths. and Physics*, Vol. 18 (1939), pp. 51–151.

Chapter 3

3.1.   J. A. LYNES AND P. BURBERRY: Geometry of Sunlight. *Architects' Jnl.* (Jan. 12th., 1966), pp. 109–131.
3.2.   H. E. BELLCHAMBERS, P. PETHERBRIDGE AND R. O. PHILLIPS: Nomenclature and symbols associated with radiation transfer calculations. *Trans. Illum. Eng. Soc. (London)*, Vol. 26 (1961), pp. 136–142.
3.3.   B. ANSTEY AND M. CHAVASSE: *The Right to Light*. The Estates Gazette, London, 1963, 113 pp.
3.4.   J. SWARBRICK: *Daylight*. Batsford, London, 1959, Chapter 2.
3.5.   A. A. GERSHUN: The Light Field. *Jnl. of Maths. and Physics*, Vol. 18 (1939), pp. 51–151.

Chapter 4

4.1.   J. C. GILCHRIST AND L. S. NESBERG: Need and perceptual change in need-related objects. *Jnl. Exp. Psychol.*, Vol. 44 (1952), pp. 369–376.
4.2.   S. M. NEWHALL, D. NICKERSON AND D. B. JUDD: Final report of the O.S.A. sub-committee on the spacing of the Munsell colors. *Jnl. of Optical Soc. of America*, Vol. 33 (1943). pp. 385–418.
4.3.   J. LONGMORE AND P. PETHERBRIDGE: Munsell value/surface reflectance relationships. *Jnl. of Optical Soc. of America.* Vol. 51 (1961), pp. 370–371.

4.4.   D. B. JUDD: Basic correlates of the visual stimulus. Chapter 22 of S. S. Stevens, *Handbook of Experimental Psychology*, Wiley, New York, 1951, pp. 811–867.

4.5.   H. R. BLACKWELL: Brightness discrimination data for the specification of quantity of illumination. *Illuminating Engineering (New York)*, Vol. 47 (1952), pp. 602–609.

4.6.   J. E. KAUFMAN (Editor): *I.E.S. Lighting Handbook* (4th Edition). Illuminating Engineering Society, New York, 1966, p. 2–14.

4.7.   H. C. WESTON: *The relation between illumination and visual efficiency—the effect of brightness contrast*. Medical Research Council Industrial Health Research Board Report No. 87, H.M. Stationery Office, London, 1945, 35 pp.

4.8.   G. A. FRY: Assessment of visual performance. *Illuminating Engineering (New York)*, Vol. 57 (1962), pp. 426–437.

4.9.   H. C. WESTON, D. J. BRIDGERS AND J. LEDGER: A study of the effect of pattern on the detection of detail at different levels of illumination. *Ergonomics*, Vol. 6 (1963), pp. 367–376.

4.10. H. HEWITT: Contribution to discussion. *Commission Internationale de l'Éclairage Compte Rendu*, 1963, p. 258.

4.11. See, for example, J. A. C. BROWN: *The Social Psychology of Industry*. Penguin Books, Harmondsworth, Middlesex, 1954, Chapter 3.

4.12. C. CUTTLE, W. B. VALENTINE, J. A. LYNES AND W. BURT: Beyond the working plane. *Commission Internationale de l'Éclairage Compte Rendu*, 1967. Paper P–67.12.

Chapter 5

5.1.   L. H. MCDERMOTT AND G. W. GORDON-SMITH: Daylight illumination recorded at Teddington. *Proc. Building Research Congress*, Division 3, Part III (1951), pp. 156–161.

5.2.   P. MOON: Proposed standard solar-radiation curves for engineering use. *Jnl. of Franklin Institute*, Vol. 230 (1940), pp. 583–617.

5.3.   D. B. JUDD, D. L. MACADAM AND G. WYSZECKI: Spectral distribution of typical daylight as a function of correlated color temperature. *Jnl. of Optical Soc. of America*, Vol. 54 (1964), pp. 1031–1040.

5.4.   J. W. T. WALSH: *The Science of Daylight*, Macdonald, London, 1961, Chapter 2.

5.5.   R. KITTLER: Standardisation of outdoor conditions for the calculation of daylight factor with clear skies. *Proc. C.I.E. Intersessional Conference on Sunlight in Buildings*, Bouwcentrum, Amsterdam, 1967, pp. 273–285.

5.6.   *Commission Internationale de l'Éclairage Compte Rendu*, 1955, Sect. 3.2, A1.

5.7.   R. KITTLER: Contribution to discussion on report of Committee E–3.2. *Commission Internationale de l'Éclairage Compte Rendu*, 1963, pp. 382–384.

5.8.   J. KROCHMANN: Über die Horizontalbeleuchtungsstärke der Tagesbeleuchtung. *Lichttechnik*, Vol. 15 (1963), pp. 559–562.

5.9.   L. A. JONES AND H. R. CONDIT: Sunlight and skylight as determinants of photographic exposure. *Jnl. of Optical Soc. of America*, Vol. 38 (1948), pp. 123–178.

5.10. G. Pleijel: *The computation of natural radiation in architecture and town planning*. Statens nämnd för Byggnadsforskning, Stockholm, 1954, 143 pp.
5.11. *Innenraumbeleuchtung mit Tageslicht*. German Standard DIN 5034, 1959.
5.12. H. H. Kimball: Sky brightness and daylight illumination measurements. *Trans. Illum. Eng. Soc.* (*New York*), Vol. 16 (1921), pp. 255–283.

Chapter 6

6.1. H. F. Stephenson: The equipment and functions of an illumination laboratory. *Trans. Illum. Eng. Soc.* (*London*), Vol. 17 (1952), pp. 1–29.
6.2. N. R. Campbell and M. K. Freeth: Compensating circuits for rectifier cells. *Jnl. of Scientific Instruments*, Vol. 11 (1934), pp. 125–126.
6.3. L. A. Wood: Zero-potential circuit for blocking-layer photo-cells. *Review of Scientific Instruments*, Vol. 7 (1936), p. 157.
6.4. R. Sewig and W. Vaillant: Die Abweichungen der Sperrschichtphotozellen vom Cosinusgesetz und eine Korrektion dafür. *Das Licht*, Vol. 4 (1934), pp. 57–58.
G. P. Barnard: The dependence of sensitivity of the selenium–sulphur rectifier photoelectric cell on the obliquity of the incident light, and a method of compensation therefor. *Proc. Physical Soc.*, Vol. 48 (1936), pp. 153–163.
G. B. Buck: Correction of light-sensitive cells for angle of incidence and spectral quality of light. *Illuminating Engineering* (*New York*), Vol. 44 (1949), pp. 293–302.
F. Hartig and H. J. Helwig: Ein cos i–gerechtes Photometer. *Lichttechnik*, Vol. 7 (1955), pp. 181–182.
O. Reeb and W. Tosberg: Ein photoelektrischer Beleuchtungsmesser hoher Empfindlichkeit mit cosinus–getreuer Bewertung. *Lichttechnik*, Vol. 7 (1955), pp. 275–278.
6.5. G. Pleijel and J. Longmore: A method of correcting the cosine error of selenium rectifier photocells. *Jnl. of Scientific Instruments*, Vol. 29 (1952), pp. 137–138.
6.6. J. A. Lynes, W. Burt, G. K. Jackson and C. Cuttle: The flow of light into buildings. *Trans. Illum. Eng. Soc.* (*London*), Vol. 31 (1966), pp. 65–91.
6.7. P. Petherbridge and W. M. Collins: The "EEL" B.R.S. Daylight Photometer. *Jnl. of Scientific Instruments*, Vol. 38 (1961), p. 375.
6.8. J. Van Den Eijk: Instrumentation for solar studies. *Proc. C.I.E. Intersessional Conference on Sunlight in Buildings*. Bouwcentrum, Amsterdam, 1967, pp. 265–272.
6.9. D. A. Schreuder: Optical system for a universal luminance meter. *Applied Optics*, Vol. 5 (1966), pp. 1965–1966.
6.10. J. B. Collins: Some applications of photography to lighting research. *Photographic Journal*, Vol. 98 (1958), pp. 218–224.
R. M. Callender: Two photographic techniques applied to lighting research. *British Jnl. of Photography*, Vol. 108 (1961), pp. 636–640.
6.11. J. W. T. Walsh: *Photometry* (3rd Edition), Constable, London, 1958, 544 pp.
6.12. H. G. W. Harding: The colour temperature of light sources. *Proc. Physical Society*, Vol. 63B (1950), pp. 689–698.
D. B. Judd: Blue-glass filters to approximate the black body at 6500°K. *Die Farbe*, Vol. 10 (1961), pp. 31–36.

6.13.   J. S. Preston and G. W. Gordon-Smith: A new determination of the luminance factor of magnesium oxide. *Proc. Physical Society*, Vol. 65B (1952), pp. 76–80.

6.14.   D. C. Croghan: The design of an artificial sky. *Architects' Jnl.* (July 22nd, 1964), pp. 215–220.

6.15.   J. Longmore: The role of models and artificial skies in daylighting design. *Trans. Illum. Eng. Soc. (London)*, Vol. 27 (1962), pp. 121–138.

Chapter 7

7.1.   C. L. Sanders: Color preferences for natural objects. *Illuminating Engineering (New York)*, Vol. 54 (1959), pp. 452–456.

H. D. Einhorn and D. M. H. Naudé: Colour-rendering preferences for lighting the face. *Trans. Illum. Eng. Soc. (London)*, Vol. 28 (1963), pp. 149–154.

7.2.   G. Reusch: Preventing the intrusion of unwelcome sunshine by means of using glass with variable transmission. *Proc. C.I.E. Intersessional Conference on Sunlight in Buildings*, Bouwcentrum, Amsterdam, 1967, pp. 199–204.

Chapter 8

8.1.   A. R. Bean and R. H. Simons: *Transmission and reflection characteristics of diffusing louvers.* Illuminating Engineering Society (London) Monograph No. 9, 1965, pp. 20–26.

8.2.   D. Paix: Daylighting with skylights. *Light and Lighting*, Vol. 55 (1962), pp. 149–151.

8.3.   K. Hisano: Light flux distribution in a rectangular parallelepiped and its simplifying scale. *Illuminating Engineering (New York)*, Vol. 41 (1946), pp. 232–247.

8.4.   *Depreciation and maintenance of interior lighting.* Illuminating Engineering Society (London) Technical Report No. 9, 1967, 31 pp.

8.5.   *Estimating daylight in buildings*—2, B.R.S. Digest (Second Series) No. 42. H.M. Stationery Office, London, 1964. 6 pp.

8.6.   R. O. Phillips: Natural lighting investigations. Appendix 2.6–1 to P. Manning, *The design of roofs for single-storey general-purpose factories.* University of Liverpool, England, 1962, pp. 140–173.

8.7.   *Lighting during daylight hours.* Illuminating Engineering Society (London) Technical Report No. 4, 1962. 31 pp.

8.8.   C. Cuttle, W. B. Valentine, J. A. Lynes and W. Burt: Beyond the working plane. *Commission Internationale de l'Éclairage Compte Rendu*, 1967. Paper P–67.12.

8.9.   R. O. Phillips and S. J. Prokhovnik: *The new approach to interreflections.* Illuminating Engineering Society (London) Monograph No. 3, 1960, 17 pp.

Chapter 9

9.1.   G. V. Pleijel: Valördiagram för beräkning av direktkvoter. *Byggmästaren*, Vol. 26 (1947), pp. 46–49.

9.2.   J. M. Waldram: The design of the visual field in streets: the visual engineer's contribution. *Trans. Illum. Eng. Soc. (London)*, Vol. 31 (1966), pp. 7–26.

H. M. Ferguson: Perspective and geometry for lighting engineers. *Light and Lighting*, Vol. 61 (1968), pp. 44–50.

9.3. J. SWARBRICK: *Daylight*, Batsford, London, 1953, Chapter 2.
9.4. H. ZIJL: *Large-size perfect diffusors* (2nd Edition), Philips Technical Library, Eindhoven, 1960, Chapter 14.
9.5. R. G. HOPKINSON, J. LONGMORE AND P. PETHERBRIDGE: An empirical formula for the computation of the indirect component of daylight factor. *Trans. Illum. Eng. Soc. (London)*, Vol. 19 (1954), pp. 201–219.
9.6. R. G. HOPKINSON, P. PETHERBRIDGE AND J. LONGMORE: *Daylighting*. Heinemann, London, 1966, pp. 522–523.
9.7. A. R. BEAN AND R. H. SIMONS: *Transmission and reflection characteristics of diffusing louvers*. Illuminating Engineering Society (London) Monograph No. 9, 1965, pp. 20–24.
9.8. P. PETHERBRIDGE: Natural lighting prediction and the design of window systems for tropical climates. *Commission Internationale de l'Éclairage Compte Rendu*, 1959, pp. 335–343.

Chapter 10

10.1. R. G. HOPKINSON AND J. LONGMORE: The permanent supplementary artificial lighting of interiors. *Trans. Illum. Eng. Soc. (London)*, Vol. 24 (1959), pp. 121–148.
10.2. T. A. MARKUS: The function of windows—a reappraisal. *Building Science*, Vol. 2 (1967), pp. 97–121.
10.3. R. G. HOPKINSON, P. PETHERBRIDGE AND J. LONGMORE: *Daylighting*. Heinemann, London, 1966, Chapters 12 and 19.
10.4. T. A. MARKUS: *The Sun and Building Design*. Elsevier, London (to be published).
10.5. L. L. HOLLADAY: Action of a light-source in the field of view in lowering visibility. *Jnl. of Optical Soc. of America*, Vol. 14 (1927), pp. 1–15.
10.6. Report of Committee W–3.1.1.2: Causes of discomfort in lighting. *Commission Internationale de l'Éclairage Compte Rendu*, 1959, pp. 249–253.
10.7. R. G. HOPKINSON AND J. LONGMORE: *Tables for glare index in daylit interiors*. National Building Studies, Special Report, H.M. Stationery Office, London (to be published). Cited in R. G. HOPKINSON, P. PETHERBRIDGE AND J. LONGMORE: *Daylighting*. Heinemann, London, 1966, p. 331.
10.8. R. G. HOPKINSON: A note on the use of indices of glare discomfort for a code of lighting. *Trans. Illum. Eng. Soc. (London)*, Vol. 25 (1960), pp. 135–138.
10.9. W. M. COLLINS: The determination of the minimum identifiable glare sensation interval using a pair-comparison method. *Trans. Illum. Eng. Soc. (London)*, Vol. 27 (1962), pp. 27–34.
10.10. *The I.E.S. Code: Recommendations for lighting building interiors*. Illuminating Engineering Society, London, 1968, 78 pp.
10.11. H. HEWITT: The study of pleasantness. *Light and Lighting*, Vol. 56 (1963), pp. 154–164.

Chapter 11

11.1. *The redevelopment of central areas*. Ministry of Town and Country Planning. H.M. Stationery Office, London, 1947, 99 pp.
11.2. D. CROGHAN: Daylight and the form of office buildings. *Architects' Jnl.* (Dec. 22nd, 1965), pp. 1501–1508.

11.3. C. T. LARSON: *The effect of windowless classrooms on elementary school children.* Architectural Research Laboratory, Department of Architecture, University of Michigan, Ann Arbor, 1965, 110 pp.

11.4. D. L. MEDD AND M. B. MEDD: Book Review—The effect of windowless classrooms on elementary school children. *Architectural Design*, Vol. 37 (1967), facing p. 152.

11.5. F. DUFFY: Bürolandschaft. *Architectural Review*, Vol. 135 (1964), pp. 148–154.

11.6. P. MANNING: *Office design: a study of environment.* Department of Building Science, University of Liverpool, England, 1965, 160 pp.

11.7. J. WEEKS: Hospitals for the 1970's. *R.I.B.A. Jnl.*, Vol. 71 (1964), pp. 507–516.

11.8. *L'éclairage naturel et ses applications.* Daylight Committee of the Belgian National Illumination Committee. Editions SIC, Brussels, 1964, p. 87.

11.9. G. THOMPSON: A new look at colour rendering, level of illumination, and protection from ultra-violet radiation in museum lighting. *Studies in Conservation*, Vol. 6 (1961), pp. 49–70.

11.10. *Éclairage et protection contre la lumière des objets colorés exposés dans les musées et galeries d'art.* Fascicule de documentation X, No. 40 – 502, Association Francaise de Normalisation (AFNOR), Paris, 1963, 6 pp.

11.11. G. THOMPSON: *Control of damage by light in the National Gallery.* Report to Ministry of Public Building and Works, National Gallery, London, 1963, 2 pp.

11.12. The lighting of art galleries. *Architects' Jnl.* (April 9th, 1959), pp. 553–558.

11.13. J. R. JOHNSON: *The radiance of Chartres*, Phaidon Press, London, 1964, 290 pp.

11.14. R. G. HOPKINSON: Daylight as a cause of glare. *Light and Lighting*, Vol. 56 (1963), pp. 322–326.

Problems

1. How many lumens are produced by 15 watts of monochromatic light at a wavelength of 520 nm?
2. What is the horizontal illumination at a point 20 feet below the centre of the dome light whose polar curve is shown in Fig. 2.3? Calculate the horizontal illumination and the scalar illumination on the same horizontal plane at a distance of 10 feet from this point.
3. Using the isolux diagram, Fig. 2.9, estimate the radius of the 5 lm/ft^2 contour on a plane 14 feet below the given dome light.
4. Find (a) the scalar illumination,
   (b) the horizontal illumination
at P due to source D shown in Figs. 2.10 and 2.11.
What illumination will it produce on a vertical surface which faces in the direction of B in plan?

Chapter 3

1. In Fig. 3.5, PE is 18 feet, AE is 13 feet, BE is 7 feet and BC is 12 feet. The vertical rectangle ABCD has a uniform luminance of 120 foot-lamberts. What is the horizontal illumination at P?
2. In Fig. 3.9, AB is 12 inches and P_1P_2 is 15 inches. If the infinite strip AB has a uniform luminance of 300 foot-lamberts, what is the average illumination along P_1P_2?

Chapter 4

1. A matt grey surface with a Munsell Value of 7 receives an illumination of 50 lumens per square foot. What is its luminance
   (a) in foot-lamberts?
   (b) in candelas per square inch?

2. What is the daylight factor 11 feet below the middle of the dome light whose polar curve is shown in Fig. 2.3?
3. The illumination vector at a chosen point indoors has a magnitude of 60 lumens per square foot and a direction 15 degrees from the downward vertical. Use Fig. 4.5 to estimate what scalar illumination would produce optimum modelling at the point.

Chapter 5

1. What is the British Standard Time at Cardiff ($4°$ W, $51\frac{1}{2}°$ N) when a correctly oriented sundial reads 10 a.m. on March 21st? (British Standard Time is one hour ahead of Greenwich Mean Time.)
Determine the height of the sun above the horizon, and the horizontal illumination on a cloudless day due to direct sunlight and due to the blue sky at this time of day.
2. For the same time and place as in the previous example what would be the mean illumination at a point indoors having a daylight factor of 1·5 per cent?
(Your answer should be based upon the mean sky illumination, excluding direct sunlight.)

Chapter 7

1. Using Table 7.1 estimate the normal transmission factor of the "grey" glass whose transmission curve is shown in Fig. 7.1 when it is illuminated by C.I.E. Illuminant D_{6500}. Use this result in conjunction with Figs. 7.3 and 7.6 to find its transmission factor at an angle of incidence of 70 degrees and also its diffuse transmission factor.

Chapter 8

1. Use eqn. (8.2) to estimate the average daylight factor in a very large room lit by 24-inch-diameter rough-cast glass dome lights on 9-inch vertical skirts finished in white paint having a reflection factor of 0·8. The glass-to-floor ratio, g/f, is 0·12. The reflection factors of floor and ceiling are 0·2 and 0·7 respectively. Ignore the effect of dirt in this example.

2. An average daylight factor of 6 per cent is required in an office 30 feet long and 20 feet wide with a floor-to-ceiling height of 9 feet. The office is situated in a clean industrial area and is to be lit by circular glass domes on white cylindrical skirts whose height is roughly half their radius. Ceiling and walls have reflection factors of 0·7 and 0·3 respectively. What area of glass will be needed?

3. A factory 400 feet long and 100 feet wide requires an average daylight factor of 8 per cent. Ceiling, walls and floor are to have reflection factors of 0·7, 0·1 and 0·1 respectively. A saw-tooth roof with sloping windows is to be used. The factory undertakes clean industrial work in a dirty urban area. Windows will be glazed with wired rough-cast glass. To take account of overhead services and glazing bars, assume that the correction factor B equals 0·6. The working plane is 3 feet above floor level; the centre of the glazing is 16 feet above the floor. What area of glazing is needed?

4. What is the light transfer ratio of the saw-tooth roof illustrated in Fig. 8.6? Use this result in conjunction with Table A.8.2, to find the coefficient of utilisation for this roof light under the following conditions:

Room index = 4·0 Reflection factors: ceiling — 0·5, walls— 0·3.

Chapter 9

1. Use Fig. A.9.1 to find the horizontal illumination at P in Chapter 3, Question 1.

2. What is the sky component of the daylight factor at a point on a horizontal plane 6 feet back from the centre of the window in Fig. 9.2 and at the same height as in Figs. 9.3 and 9.4?

3. Estimate the sky component and the externally reflected component for the window shown in Fig. 9.7. All dimensions and distances are as in Fig. 9.7, but the reference point is moved sideways to a distance equal to half the width of the glazing, i.e., it is now opposite the corner of the window. Assume very thin walls.

4. Find the magnitude and direction of the illumination vector at the same position as in Question 3.

5. Find the total daylight factor, in scalar terms, at the reference point in Fig. 9.7. Assume that the room, a laboratory in a dirty industrial area, is 12 feet square and 8 ft 6 in high; the glazed area of the window is 6 ft by 12 ft 6 in. The reference point, level with the middle of the window sill, is 2 ft 9 in above the floor. The reflection factors are: ceiling — 0·7, walls — 0·5, floor — 0·1. The roof-line facing the window has an angle of elevation of 20 degrees from the centre of the window.

Chapter 10

1. Estimate the Glare Index on an overcast day at the reference point in Fig. 9.7, for an observer facing the window, when the outdoor illumination is 500 lm/ft^2, and when it is 2000 lm/ft^2. Assume that $(ERC_{vert} + IRC) = 1$ per cent.

Answers to Problems

Chapter 2

1. 7250 lumens
2. 6·6 lm/ft², 3·2 lm/ft², 3·6 lm/ft².
3. 8·4 ft.
4. 7·5 lm/ft², 21 lm/ft², 12·8 lm/ft².

Chapter 3

1. 5.5 lm/ft².
2. 160 lm/ft².

Chapter 4

1. 21 ft-L, 0·042 cd/in².
2. 2.2 per cent.
3. 40 lm/ft².

Chapter 5

1. 23½ minutes past 11, 33 degrees, 4600 lm/ft², 1000 lm/ft².
2. 26 lm/ft².

Chapter 7

1. 0·45, 0·28, 0·37.

Chapter 8

1. 8 per cent.
2. 100 ft².
3. 18600 ft², 3·5 per cent.
4. 0·51, 0·35.

Chapter 9

1. 5.5 lm/ft^2.

2. 1·3 per cent.

3. 0·46 per cent, 0·08 per cent.

4. 1·6 per cent, 18$\frac{1}{2}$ degrees below horizontal, inclined at 23$\frac{1}{2}$ degrees on plan from the normal to the window.

5. 0·79 per cent.

Chapter 10

1. 20·4, 24·0.

Appendix

Note: The numbering of Tables and Charts in this Appendix is related to the numbering of the relevant chapters in the text.

Table A.2.I

Units of Illumination

Unit	Name	Expressed in lm/ft^2
lumen per square foot	foot-candle (f.c.)	1 lm/ft^2
lumen per square metre	lux (lx)	0.0929 lm/ft^2
lumen per square centimetre	phot (ph)	929 lm/ft^2
10^{-3} phot	milliphot	0.929 lm/ft^2
10^{-3} lux	nox	$0.0000929 \text{ lm/ft}^2$

Table A.2.II

*Illumination Calculation Data**

ϕ	$\cos^3 \phi$	$0.25 \cos^2 \phi$	ϕ	$\cos^3 \phi$	$0.25 \cos^2 \phi$
$0°$	1·0000	0·2500	$50°$	0·2656	0·1033
$5°$	0·9886	0·2481	$55°$	0·1887	0·0822
$10°$	0·9551	0·2425	$60°$	0·1250	0·0625
$15°$	0·9012	0·2333	$65°$	0·0755	0·0447
$20°$	0·8298	0·2208	$70°$	0·0400	0·0292
$25°$	0·7444	0·2053	$75°$	0·0173	0·0167
$30°$	0·6495	0·1875	$80°$	0·00524	0·00754
$35°$	0·5497	0·1678	$85°$	0·000662	0·00190
$40°$	0·4495	0·1467	$90°$	0	0
$45°$	0·3536	0·1250			

* To be used in conjunction with eqns. (2.12) and (2.13) (*see* Fig. 2.7).

$$\text{Horizontal illumination} = \frac{I(\phi)}{h^2} \times \cos^3 \phi.$$

$$\text{Scalar illumination} = \frac{I(\phi)}{h^2} \times 0.25 \cos^2 \phi.$$

177

Table A.3

Units of Luminance

Unit	Name	Expressed in ft-L
lm/ft^2 of uniform diffuser*	foot-lambert (ft-L) or equivalent foot-candle (e.f.c.)	1 ft-L
lm/m^2 of uniform diffuser*	apostilb (asb), blondel or equivalent lux	0·0929 ft-L
lm/cm^2 of uniform diffuser*	lambert, or equivalent phot	929 ft-L
10^{-3} lamberts	millilambert (mL)	0·929 ft-L
10^{-3} apostilbs	skot	0·0000929 ft-L
cd/ft^2	candela per square foot (cd/ft^2)	3·1416 ft-L
cd/in^2	candela per square inch (cd/in^2)	452·4 ft-L
cd/m^2	nit	0·2919 ft-L
cd/cm^2	stilb (sb)	2919 ft-L

* See Section 3.2.

Table A.8.I

Coefficients of Utilisation for Glass Domes

				Reflection factor					
Ceiling		0·7			0·5			0·3	0
Walls	0·5	0·3	0·1	0·5	0·3	0·1	0·3	0·1	0
Room index				Coefficients of utilisation					
Skirt 0·6	0·43	0·38	0·34	0·42	0·38	0·34	0·37	0·34	0·33
height=0. 0·8	0·52	0·47	0·43	0·51	0·46	0·43	0·46	0·43	0·41
S/H=1·25. 1·0	0·56	0·51	0·47	0·55	0·51	0·47	0·50	0·47	0·45
1·25	0·62	0·56	0·53	0·60	0·56	0·52	0·55	0·52	0·50
1·5	0·65	0·60	0·56	0·64	0·59	0·56	0·58	0·55	0·54
2·0	0·70	0·66	0·62	0·69	0·65	0·62	0·64	0·61	0·60
2·5	0·74	0·70	0·67	0·73	0·69	0·66	0·68	0·65	0·64
3·0	0·77	0·74	0·70	0·75	0·72	0·70	0·71	0·69	0·67
4·0	0·80	0·77	0·75	0·78	0·76	0·73	0·74	0·73	0·71
5·0	0·82	0·79	0·77	0·80	0·78	0·76	0·77	0·75	0·73
Inf.	0·90	0·90	0·90	0·88	0·88	0·88	0·86	0·86	0·84

Table A.8.I—continued

Skirt	0·6	0·33	0·29	0·27	0·32	0·29	0·26	0·28	0·26	0·25
height=	0·8	0·40	0·36	0·33	0·39	0·35	0·33	0·35	0·33	0·31
½×skirt	1·0	0·44	0·41	0·38	0·43	0·40	0·38	0·40	0·38	0·36
radius.	1·25	0·48	0·45	0·42	0·48	0·44	0·42	0·44	0·42	0·40
$S/H=1·0$.	1·5	0·51	0·48	0·45	0·50	0·47	0·45	0·47	0·45	0·44
	2·0	0·55	0·52	0·47	0·54	0·51	0·47	0·50	0·48	0·47
	2·5	0·57	0·54	0·52	0·56	0·54	0·51	0·53	0·51	0·50
	3·0	0·59	0·55	0·54	0·58	0·55	0·54	0·55	0·53	0·52
	4·0	0·60	0·59	0·57	0·59	0·58	0·56	0·57	0·56	0·54
	5·0	0·62	0·60	0·59	0·61	0·60	0·58	0·58	0·57	0·56
	Inf.	0·68	0·68	0·68	0·66	0·66	0·66	0·65	0·65	0·63
Skirt	0·6	0·28	0·26	0·24	0·28	0·26	0·24	0·25	0·24	0·23
height=	0·8	0·33	0·31	0·28	0·33	0·30	0·28	0·30	0·28	0·27
1×skirt	1·0	0·38	0·34	0·32	0·36	0·34	0·32	0·34	0·32	0·31
radius.	1·25	0·40	0·37	0·35	0·39	0·36	0·35	0·36	0·35	0·34
$S/H=1·0$.	1·5	0·42	0·40	0·38	0·41	0·39	0·37	0·39	0·37	0·36
	2·0	0·44	0·42	0·40	0·43	0·41	0·40	0·41	0·40	0·39
	2·5	0·46	0·44	0·42	0·45	0·43	0·42	0·43	0·42	0·41
	3·0	0·47	0·46	0·45	0·46	0·45	0·44	0·44	0·43	0·42
	4·0	0·48	0·47	0·46	0·47	0·46	0·45	0·46	0·45	0·44
	5·0	0·49	0·48	0·47	0·48	0·48	0·47	0·47	0·46	0·45
	Inf.	0·53	0·53	0·53	0·52	0·52	0·52	0·51	0·51	0·49
Skirt	0·6	0·19	0·17	0·15	0·18	0·17	0·15	0·17	0·15	0·15
height=	0·8	0·23	0·22	0·20	0·23	0·21	0·20	0·21	0·20	0·20
2×skirt	1·0	0·25	0·24	0·23	0·25	0·24	0·23	0·23	0·22	0·22
radius.	1·25	0·27	0·25	0·24	0·27	0·25	0·24	0·25	0·24	0·24
$S/H=0·75$.	1·5	0·28	0·27	0·26	0·28	0·27	0·26	0·26	0·25	0·25
	2·0	0·30	0·29	0·28	0·29	0·28	0·28	0·28	0·27	0·27
	2·5	0·31	0·30	0·29	0·30	0·29	0·29	0·29	0·28	0·28
	3·0	0·32	0·30	0·30	0·31	0·30	0·29	0·29	0·29	0·28
	4·0	0·32	0·32	0·31	0·32	0·31	0·30	0·31	0·30	0·30
	5·0	0·33	0·32	0·31	0·32	0·32	0·31	0·31	0·30	0·30
	Inf.	0·35	0·35	0·35	0·35	0·35	0·35	0·34	0·34	0·33
Skirt	0·6	0·11	0·10	0·10	0·11	0·10	0·10	0·10	0·10	0·10
height=	0·8	0·13	0·12	0·11	0·12	0·12	0·11	0·12	0·11	0·11
4×skirt	1·0	0·13	0·13	0·12	0·13	0·13	0·12	0·12	0·12	0·12
radius.	1·25	0·14	0·13	0·13	0·14	0·13	0·13	0·13	0·13	0·13
$S/H=0·5$	1·5	0·15	0·14	0·13	0·15	0·14	0·13	0·14	0·13	0·13
	2·0	0·15	0·15	0·14	0·15	0·14	0·14	0·14	0·14	0·14
	2·5	0·16	0·15	0·15	0·16	0·15	0·15	0·15	0·15	0·14
	3·0	0·16	0·16	0·15	0·16	0·16	0·15	0·15	0·15	0·15
	4·0	0·17	0·16	0·16	0·16	0·16	0·16	0·16	0·15	0·15
	5·0	0·17	0·16	0·16	0·17	0·16	0·16	0·16	0·16	0·15
	Inf.	0·18	0·18	0·18	0·18	0·18	0·18	0·17	0·17	0·17

Table A.8.II

Room Index	Reflection factor								
Ceiling	0·7			0·5			0·3		0
Walls	0·5	0·3	0·1	0·5	0·3	0·1	0·3	0·1	0
	Coefficient of utilisation								
0·6	0·34	0·30	0·27	0·34	0·30	0·27	0·30	0·27	0·27
0·8	0·40	0·39	0·36	0·40	0·39	0·36	0·39	0·36	0·35
1·0	0·45	0·43	0·41	0·44	0·42	0·41	0·42	0·41	0·38
1·25	0·50	0·47	0·46	0·50	0·47	0·45	0·47	0·45	0·44
1·5	0·51	0·49	0·47	0·51	0·49	0·47	0·49	0·46	0·46
2·0	0·57	0·55	0·53	0·56	0·53	0·52	0·53	0·52	0·51
2·5	0·59	0·56	0·55	0·59	0·56	0·55	0·55	0·53	0·53
3·0	0·62	0·60	0·59	0·62	0·59	0·58	0·59	0·58	0·56
4·0	0·64	0·63	0·61	0·64	0·63	0·61	0·61	0·60	0·60
5·0	0·68	0·65	0·65	0·66	0·65	0·63	0·63	0·62	0·62
Inf.	0·76	0·76	0·76	0·74	0·74	0·74	0·73	0·73	0·71
0·6	0·07	0·06	0·04	0·07	0·05	0·04	0·05	0·03	0·03
0·8	0·11	0·08	0·07	0·10	0·08	0·06	0·08	0·06	0·05
1·0	0·14	0·11	0·10	0·13	0·10	0·09	0·10	0·08	0·07
1·25	0·16	0·13	0·12	0·15	0·13	0·11	0·12	0·10	0·09
1·5	0·17	0·15	0·13	0·16	0·14	0·12	0·13	0·12	0·10
2·0	0·19	0·17	0·16	0·18	0·16	0·15	0·15	0·14	0·12
2·5	0·21	0·20	0·18	0·20	0·18	0·17	0·17	0·16	0·14
3·0	0·22	0·21	0·19	0·21	0·19	0·18	0·18	0·17	0·15
4·0	0·24	0·22	0·21	0·22	0·21	0·20	0·19	0·18	0·17
5·0	0·25	0·24	0·23	0·23	0·22	0·21	0·20	0·20	0·18

Shed roof
Spacing (*see* Fig. 8.3)
LTR = 0·94

Saw-tooth with vertical window
S/H = 2:1
LTR = 0·34

Saw-tooth with sloping window
$S/H = 2{:}1$
LTR $= 0.58$

0·8	0·25	0·21	0·20	0·25	0·21	0·20	0·21	0·20	0·18
1·0	0·30	0·26	0·25	0·29	0·26	0·24	0·25	0·24	0·21
1·25	0·31	0·30	0·27	0·31	0·29	0·26	0·27	0·26	0·24
1·5	0·34	0·31	0·30	0·32	0·31	0·29	0·30	0·27	0·26
2·0	0·36	0·35	0·32	0·36	0·34	0·32	0·34	0·32	0·29
2·5	0·39	0·38	0·35	0·38	0·36	0·34	0·35	0·32	0·31
3·0	0·40	0·39	0·38	0·40	0·38	0·36	0·36	0·35	0·32
4·0	0·42	0·41	0·40	0·41	0·40	0·39	0·39	0·38	0·35
5·0	0·44	0·42	0·41	0·42	0·41	0·40	0·40	0·39	0·36
Inf.	0·49	0·49	0·49	0·48	0·48	0·48	0·45	0·45	0·42

$S/H = 2{:}1$
LTR $= 0.31$

0·6	0·07	0·05	0·04	0·06	0·05	0·04	0·05	0·04	0·03
0·8	0·09	0·07	0·06	0·09	0·07	0·06	0·07	0·06	0·05
1·0	0·12	0·10	0·08	0·11	0·09	0·08	0·09	0·08	0·07
1·25	0·14	0·12	0·10	0·13	0·11	0·10	0·11	0·10	0·09
1·5	0·15	0·13	0·12	0·15	0·13	0·12	0·13	0·11	0·11
2·0	0·17	0·15	0·14	0·16	0·15	0·14	0·15	0·13	0·13
2·5	0·18	0·17	0·15	0·18	0·16	0·15	0·16	0·15	0·14
3·0	0·20	0·18	0·17	0·19	0·18	0·17	0·17	0·16	0·16
4·0	0·21	0·20	0·19	0·20	0·19	0·19	0·19	0·18	0·17
5·0	0·22	0·21	0·20	0·21	0·20	0·19	0·20	0·19	0·18
Inf.	0·25	0·25	0·25	0·25	0·25	0·25	0·24	0·24	0·23

$S/H = 2{:}1$
LTR $= 0.30$

0·6	0·07	0·05	0·04	0·07	0·05	0·04	0·05	0·04	0·04
0·8	0·10	0·08	0·06	0·10	0·07	0·06	0·07	0·06	0·06
1·0	0·13	0·11	0·08	0·12	0·11	0·08	0·11	0·08	0·08
1·25	0·16	0·13	0·11	0·14	0·13	0·11	0·13	0·11	0·10
1·5	0·17	0·14	0·12	0·16	0·14	0·12	0·13	0·12	0·12
2·0	0·19	0·17	0·16	0·18	0·17	0·16	0·16	0·14	0·14
2·5	0·20	0·18	0·17	0·19	0·18	0·17	0·18	0·17	0·16
3·0	0·22	0·19	0·18	0·20	0·19	0·18	0·19	0·18	0·17
4·0	0·23	0·22	0·20	0·23	0·22	0·20	0·20	0·20	0·19
5·0	0·24	0·23	0·22	0·24	0·23	0·22	0·22	0·22	0·20
Inf.	0·29	0·29	0·29	0·27	0·27	0·27	0·27	0·27	0·26

Table A.8.II—continued

Ceiling	0·7			0·5			0·3		0
Walls	0·5	0·3	0·1	0·5	0·3	0·1	0·3	0·1	0
0·6	0·17	0·14	0·12	0·17	0·14	0·12	0·13	0·12	0·12
0·8	0·23	0·19	0·17	0·23	0·19	0·17	0·19	0·17	0·16
1·0	0·27	0·24	0·22	0·26	0·24	0·22	0·23	0·22	0·19
1·25	0·30	0·26	0·24	0·29	0·26	0·24	0·26	0·24	0·23
1·5	0·32	0·29	0·26	0·31	0·29	0·26	0·27	0·26	0·24
2·0	0·35	0·31	0·30	0·34	0·31	0·29	0·31	0·29	0·27
2·5	0·37	0·35	0·32	0·35	0·34	0·31	0·32	0·31	0·30
3·0	0·40	0·36	0·35	0·38	0·36	0·34	0·35	0·34	0·32
4·0	0·41	0·38	0·37	0·40	0·38	0·36	0·37	0·36	0·35
5·0	0·42	0·41	0·38	0·42	0·40	0·38	0·38	0·37	0·36
Inf.	0·49	0·49	0·49	0·48	0·48	0·48	0·47	0·47	0·46
0·6	0·15	0·12	0·09	0·13	0·12	0·09	0·11	0·09	0·09
0·8	0·19	0·16	0·13	0·19	0·16	0·13	0·16	0·13	0·13
1·0	0·23	0·20	0·18	0·22	0·19	0·18	0·19	0·18	0·16
1·25	0·26	0·23	0·20	0·24	0·23	0·20	0·22	0·20	0·19
1·5	0·27	0·24	0·22	0·26	0·24	0·22	0·23	0·22	0·20
2·0	0·30	0·27	0·24	0·28	0·26	0·24	0·26	0·24	0·24
2·5	0·32	0·30	0·27	0·31	0·28	0·27	0·28	0·27	0·26
3·0	0·34	0·31	0·30	0·32	0·31	0·28	0·30	0·28	0·27
4·0	0·35	0·34	0·32	0·34	0·32	0·31	0·32	0·31	0·30
5·0	0·35	0·34	0·34	0·35	0·34	0·32	0·34	0·32	0·31
Inf.	0·40	0·40	0·40	0·40	0·40	0·40	0·39	0·39	0·38

$S/H = 2{:}1$
LTR $= 0{\cdot}59$

$S/H = 2{:}1$
LTR $= 0{\cdot}46$

Notes
1. Values of S/H represent the maximum recommended spacing/height ratio for each type of roof light.
2. For monitor roof lights the spacing S is measured between adjacent monitors, not between adjacent windows.
3. The height H is measured from the working plane to the centre-line of the windows.

Table A.8.III

Maintenance Factors for Windows and Roof Lights [8.4.]

Location of building	Inclination of glazing	Non-industrial or clean industrial work	Dirty industrial work
Non-industrial or clean industrial area	Vertical	0·9	0·8
	Sloping	0·8	0·7
	Horizontal	0·7	0·6
Dirty industrial area	Vertical	0·8	0·7
	Sloping	0·7	0·6
	Horizontal	0·6	0·5

Table A.8.IV

Values of G [8.5]

	Vertically glazed sawtooth and monitor roofs, and side windows	All other windows
Single glazing		
Sheet or rough-cast glass	1·0	1·1
Wired cast	0·9	1·0
Diffusing opal ⅛ in acrylic plastics (depending on grade)	0·65–0·9	0·75–1·0
Corrugated glass fibre reinforced sheets:		
Moderately diffusing	0·9	1·0
Heavily diffusing	0·75–0·9	0·9 –1·0
Very heavily diffusing	0·65–0·8	0·75–0·9
Double glazing		
Rough-cast (2 sheets)	0·85	1·0
Rough-cast/wired cast	0·75	0·85

Table A.10

Recommended Minimum Daylight Factors (based upon the 1968 *I.E.S. Code* [10.10])

Note: *Except where otherwise stated these values are expressed in terms of the daylight factor on a horizontal working plane.*

The following daylight factor recommendations ensure that when a building is lighted by daylight alone, the illumination values over the working area will meet the requirements of the occupation during most daytime working hours and the environment will be acceptably comfortable. The recommended values agree generally with those given in *British Standard Code of Practice* CP 3, Chapter I, Part 1 (1964): "Daylighting".

(1) *Daylight factors for buildings with top lighting*

In factories and other buildings with roof lighting, the daylight factor should not be less than 5 per cent.

(2) *Daylight factors for buildings with side lighting*

In buildings with side lighting only, the daylight factor at points remote from the window should not be less than the value given in the Table below for the location. These recommended minimum daylight factors ensure that the building will appear to be well lighted provided that the room surfaces have suitable reflectances.

(3) *Daylight factors for buildings with part side and part top lighting*

In buildings which are lighted either from clerestory windows or partly from roof lights and partly from side windows, the daylight factors should have values intermediate between those for side lighted and top lighted buildings.

Table A.10—*continued*

		Recommended minimum daylight factor per cent
GENERAL BUILDING AREAS		
Entrance halls and reception areas		1
OFFICES		
Offices	General	2
Drawing Offices	General	2
Banks	Counters, typing, accounting, book areas	2
	Public areas	2
Telephone Exchanges (Manual)	General	2
Airport Buildings and Coach Stations	Reception areas (desks), customs and immigration halls	2
	Circulation areas, lounges	1
Assembly and Concert Halls	Foyers, auditoria	1
	Corridors	0·5
	Stairs	1
Churches	Body of church	1
	Pulpit and lectern areas, chancel, choir	1·5
	Altar, communion table	3–6[1]
	Vestries	2
	[1]Depending on emphasis required	
Libraries	Shelves (stacks)	1[1]
	Reading tables	1
	[1]Additional artificial lighting will be required	
Museums and Art Galleries	General	1[1]
	[1]Attention should be paid to conservation requirements	
Schools and Colleges	Assembly halls	2
	Classrooms	2
	Art rooms	4
	Laboratories (benches)	3
	Staff rooms, common rooms	1
Hospitals	Reception and waiting rooms	2
	Wards	1
	Pharmacies	3

APPENDIX

Table A.10—*continued*

Surgeries	Waiting rooms	2
(Medical and	Surgeries	2
dental)	Laboratories	3
Sports Halls	General	2
Swimming Pools	Surface of pool	2
	Surrounding areas	1

HOMES AND HOTELS

(In homes and rooms where a domestic atmosphere is appropriate, uniform illumination is undesirable. Pools of light so placed that the room can easily be furnished and readily adapted for all probable uses are needed.)

Living rooms: A daylight factor of not less than 1 per cent should be provided over at least 8 m² and should extend to at least half the depth of the room from the main window.

Bedrooms: A daylight factor of not less than 0·5 per cent should be provided over at least 6 m² and should extend to at least half the depth of the room from the main window.

Kitchens: A daylight factor of not less than 2 per cent should be provided over not less than 5 m² or over 50 per cent of the total floor area.

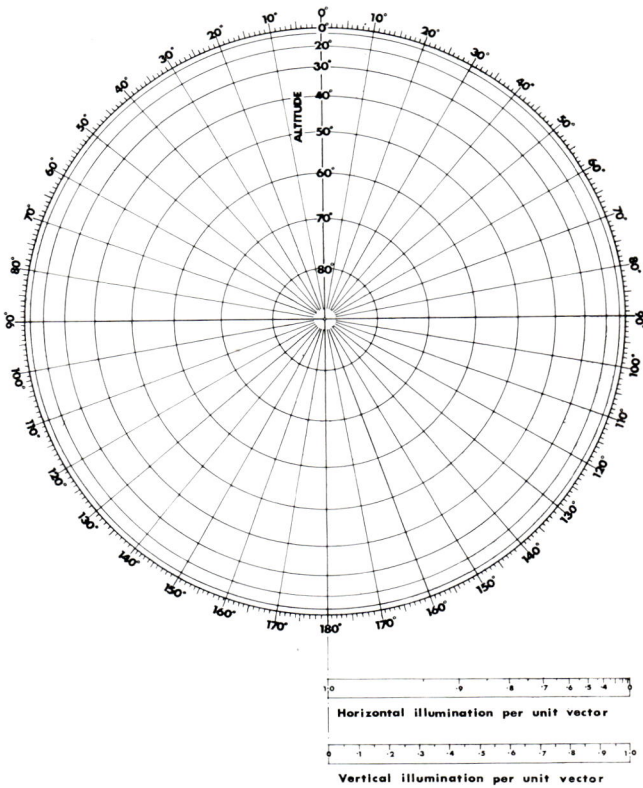

Fig. A.2.   Orthographic grid. Radius of unit hemisphere = 4 cm.

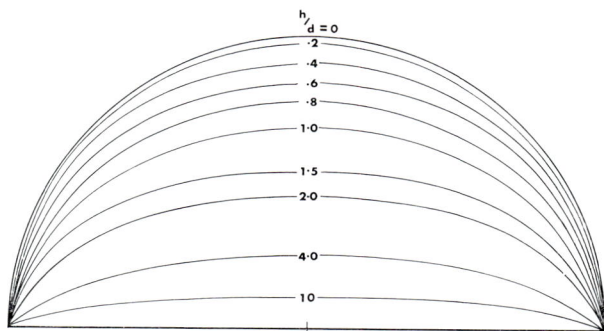

Fig. A.3.   Guide-lines for orthographic grid (*see* Section 3.2).

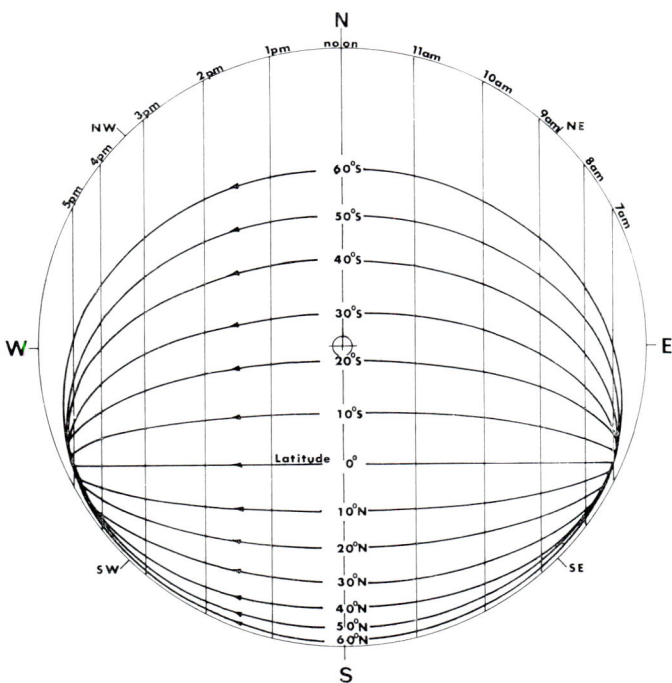

Fig. A.5.1.   Solar orbits on unit hemisphere; December 22.

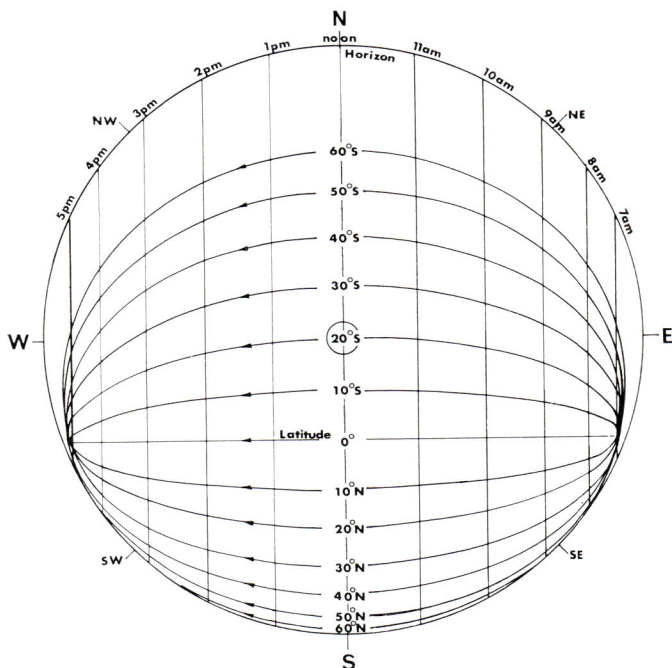

Fig. A.5.2.   Solar orbits on unit hemisphere; January 21 and November 22.

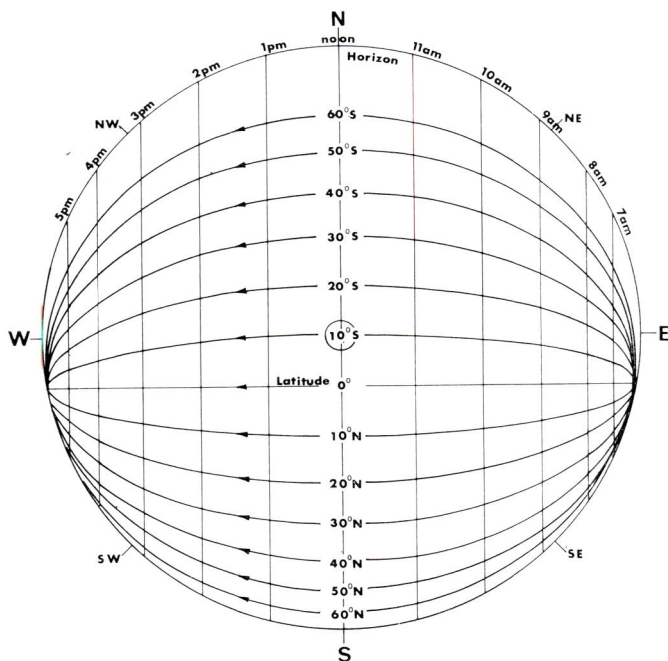

Fig. A.5.3.   Solar orbits on unit hemisphere; February 23 and October 30.

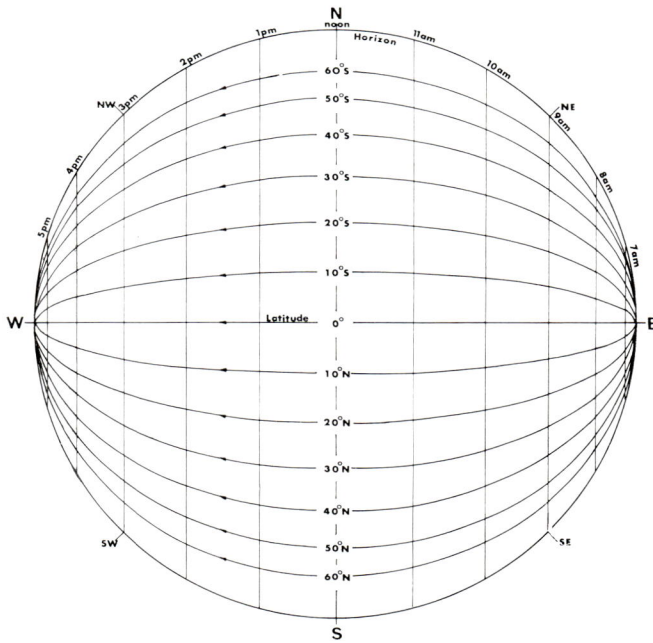

Fig. A.5.4.   Solar orbits on unit hemisphere; March 21 and September 23.

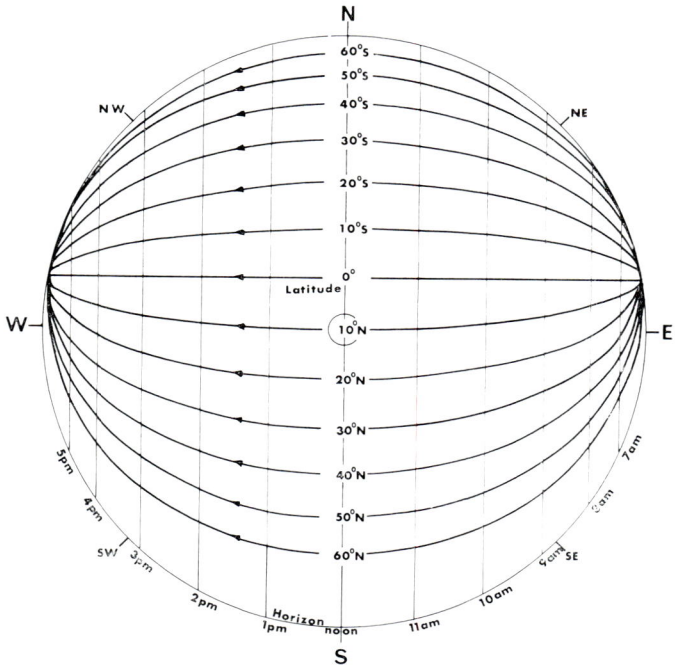

*Fig.*A. 5.5.   Solar orbits on unit hemisphere; April 16 and August 28.

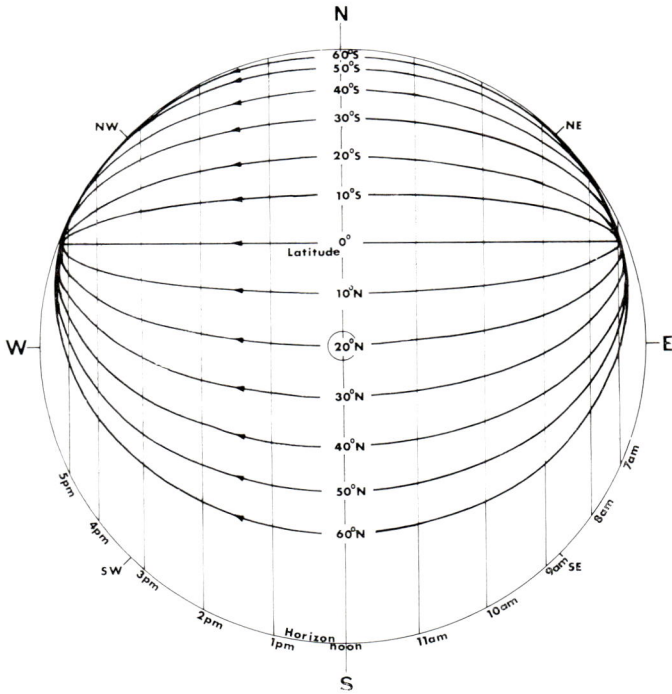

Fig. A.5.6.   Solar orbits on unit hemisphere; May 21 and July 24.

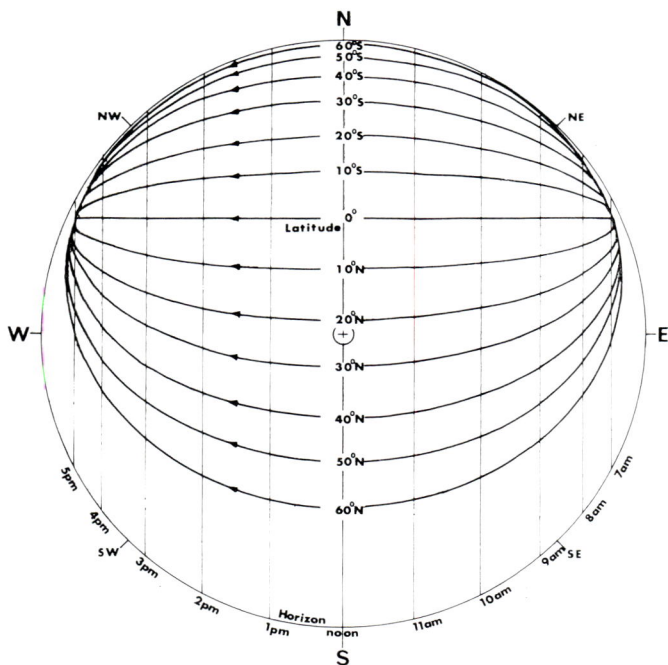

Fig. A.5.7.   Solar orbits on unit hemisphere; June 22.

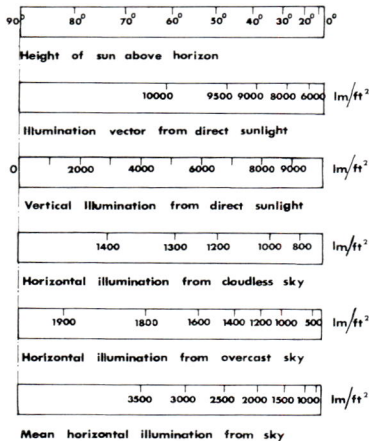

Fig. A.5.8.   Radial scales for use with solar orbits (*see* Sections 5.2 and 5.4).

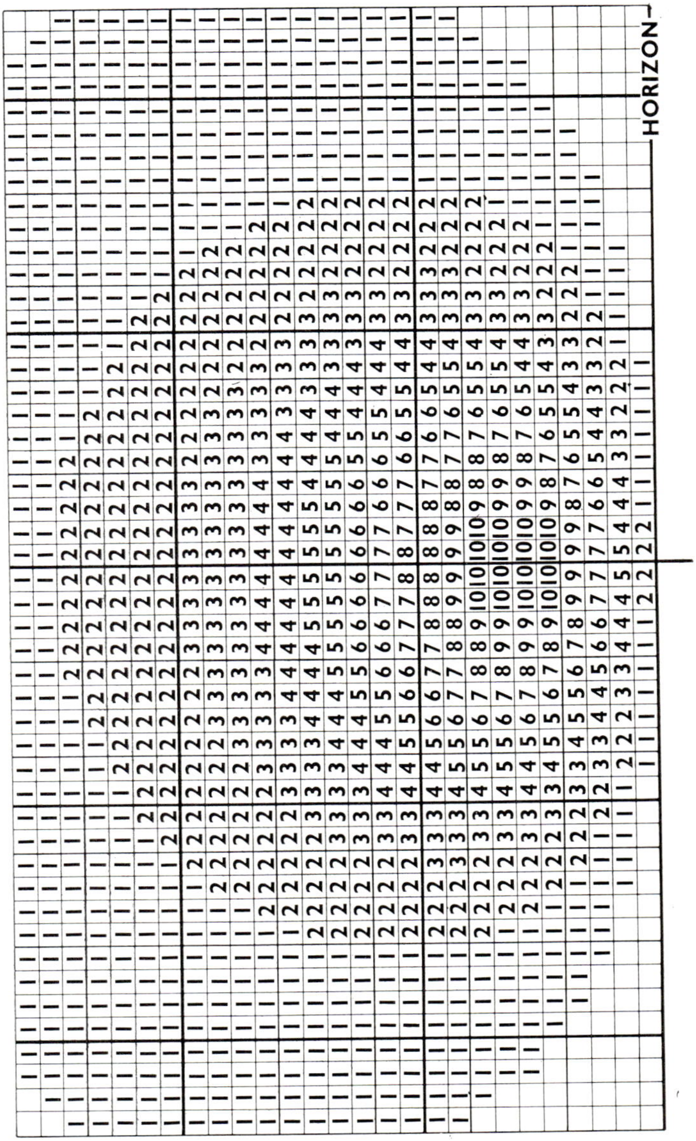

Fig. A.9.1.   One-eighth-inch graticule for sky factor on horizontal working plane, and for configuration factors.

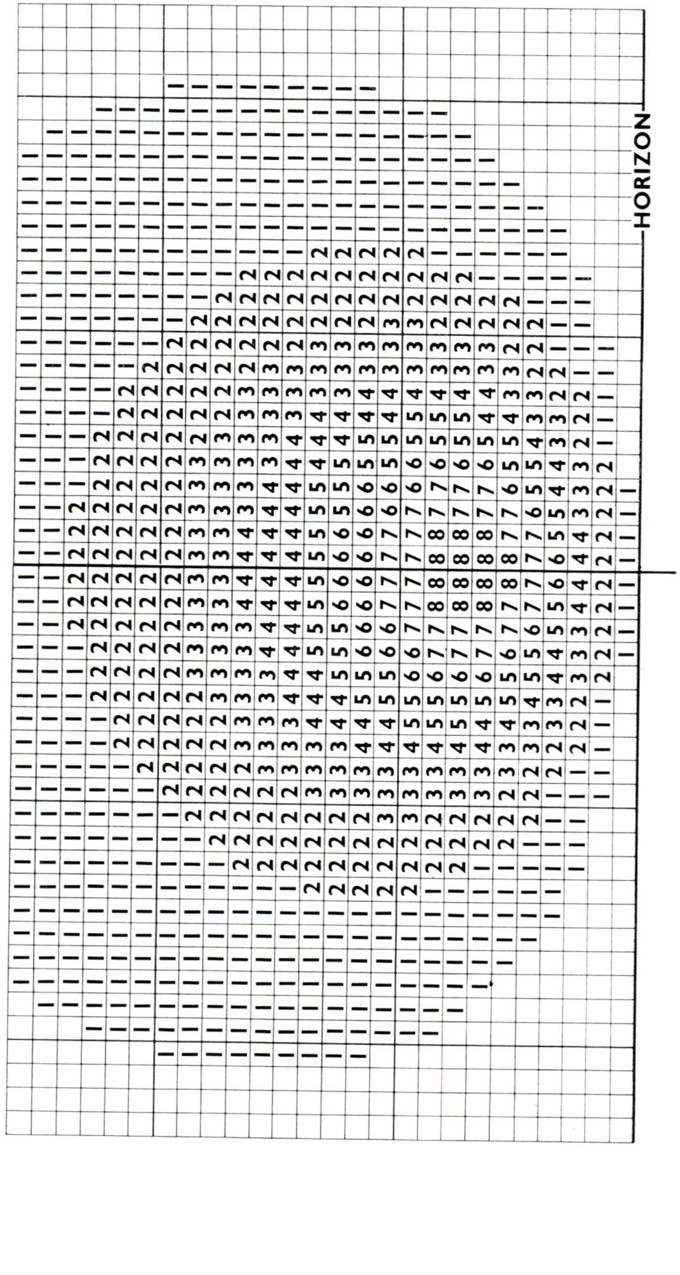

APPENDIX

Fig. A.9.2.   One-eighth-inch graticule for sky component on horizontal working plane, incorporating corrections for clear transparent glazing and C.I.E. Standard Overcast Sky.

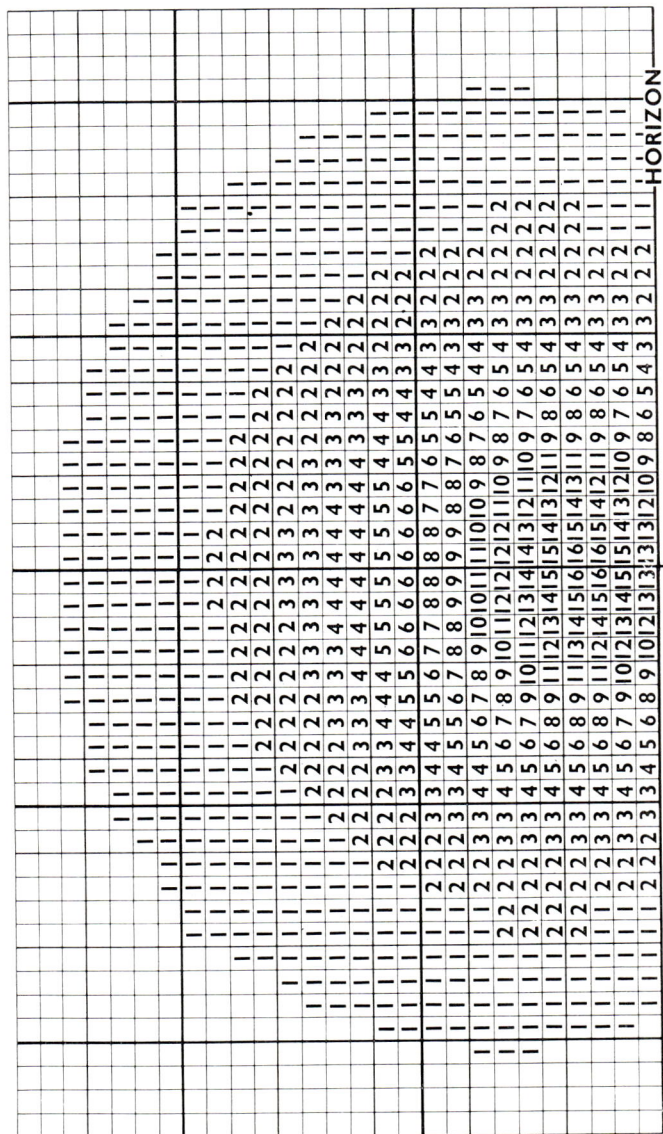

Fig. A.9.3.   One-eighth-inch graticule for sky component on vertical plane parallel to window wall, incorporating corrections for clear transparent glazing and C.I.E. Standard Overcast Sky.

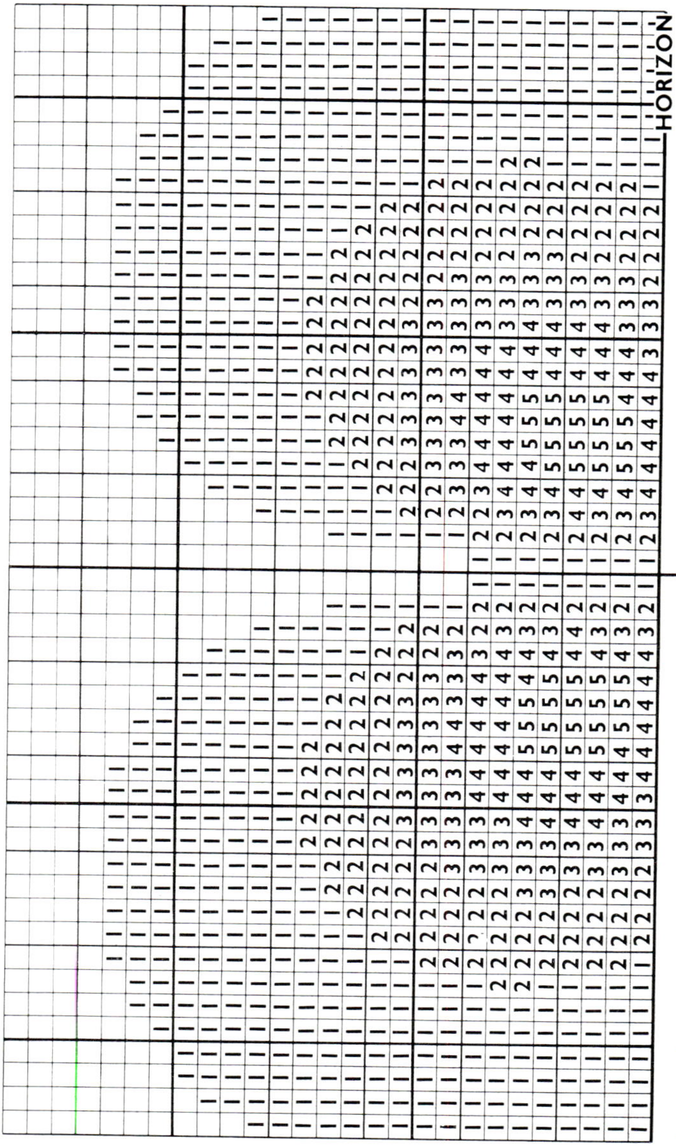

Fig. A.9.4. One-eighth-inch graticule for sky component on vertical plane perpendicular to window wall, incorporating corrections for clear transparent glazing and C.I.E. Standard Overcast Sky.

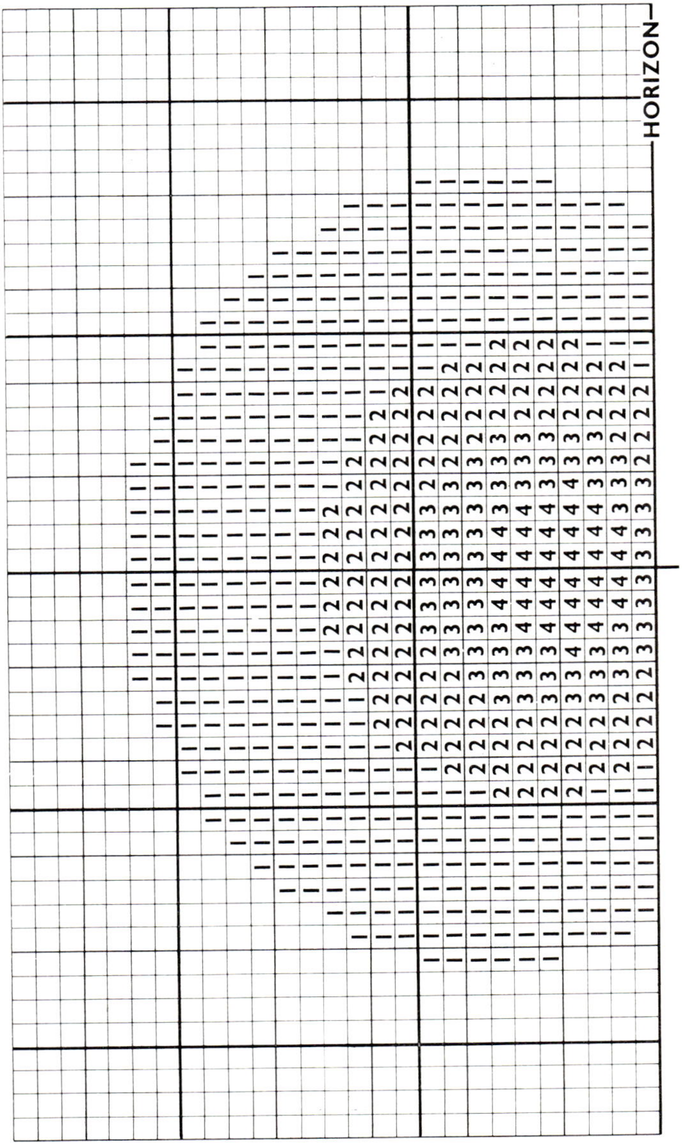

Fig. A.9.5.   One-eighth-inch graticule for scalar sky component.

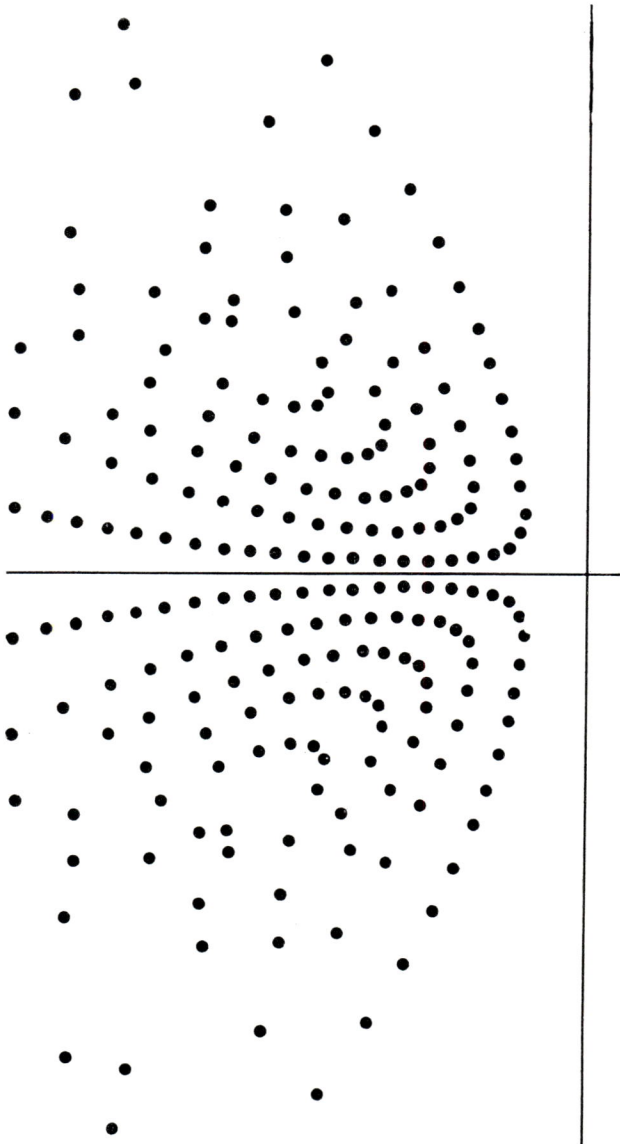

Fig. A.9.6. Pepper-pot diagram for sky component on horizontal working plane, incorporating corrections for clear transparent glazing and C.I.E. Standard Overcast Sky. The horizontal line represents the horizon.

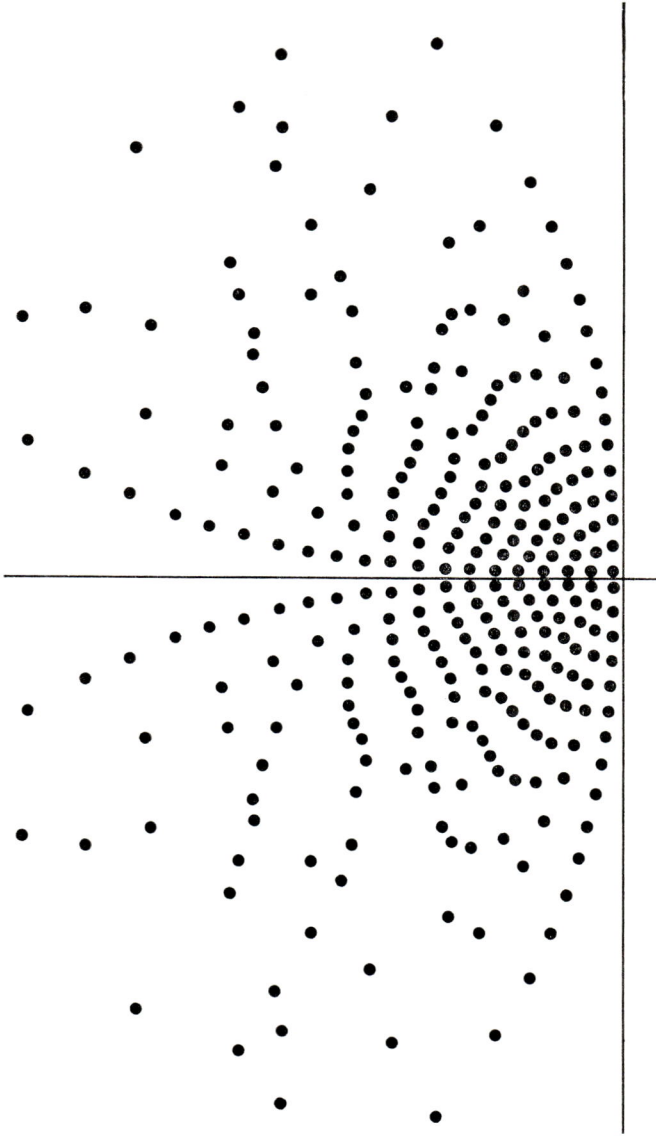

Fig. A.9.7.   Pepper-pot diagram for sky component on vertical plane parallel to window wall, incorporating corrections for clear transparent glazing and C.I.E. Standard Overcast Sky. The horizontal line represents the horizon.

APPENDIX

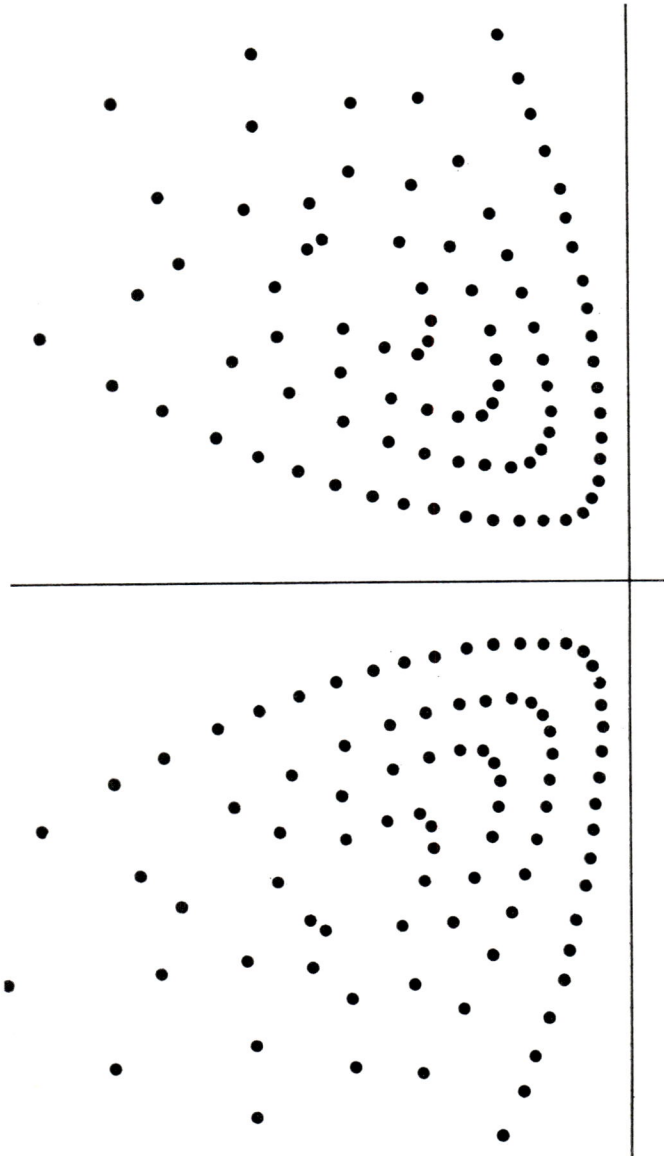

Fig. A.9.8. Pepper-pot diagram for sky component on vertical plane perpendicular to window wall, incorporating corrections for clear transparent glazing and C.I.E. Standard Overcast Sky. The horizontal line represents the horizon.

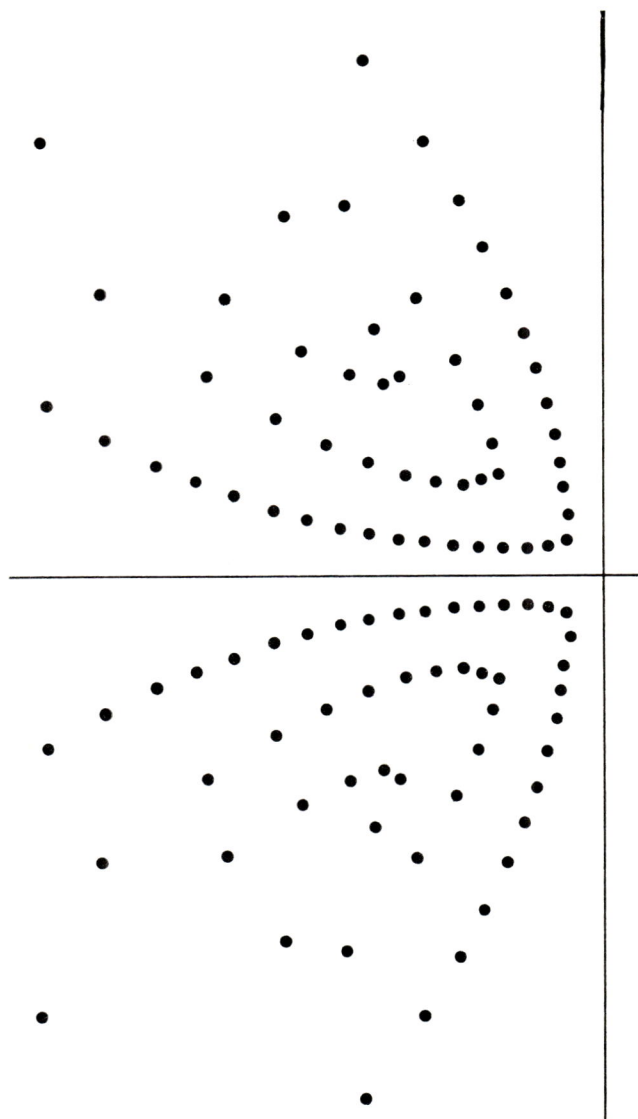

Fig. A.9.9. Pepper-pot diagram for scalar sky component, incorporating corrections for clear transparent glazing and C.I.E. Standard Overcast Sky. The horizontal line represents the horizon.

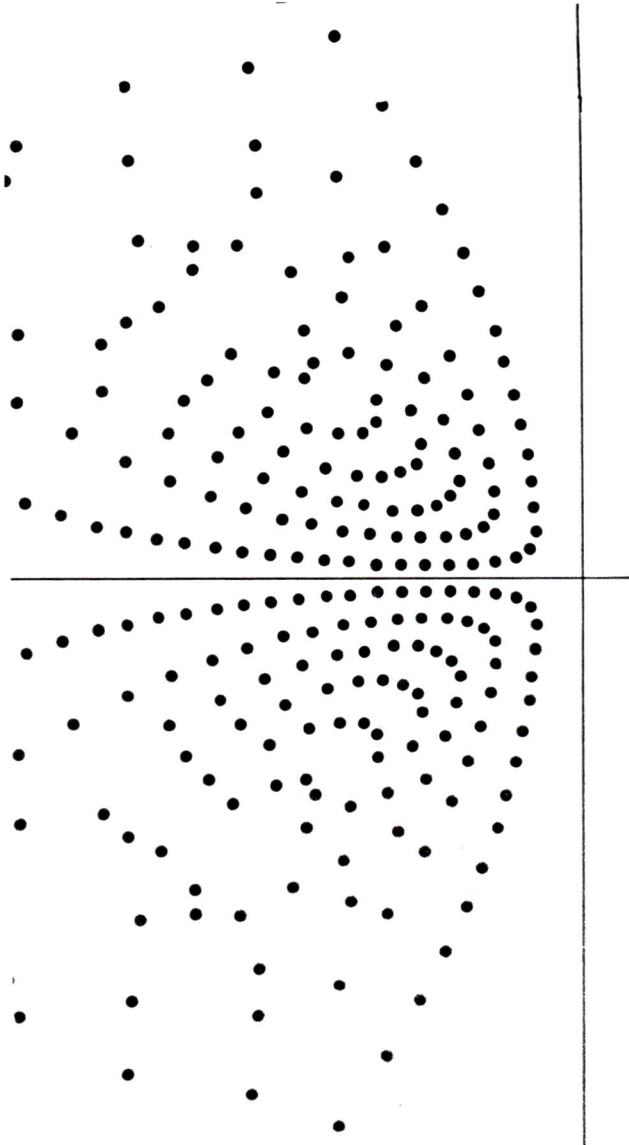

Fig. A.9.10. Pepper-pot diagram for sky component on horizontal working plane, based on sky of uniform luminance, but incorporating a correction for transparent glazing. The horizontal line represents the horizon.

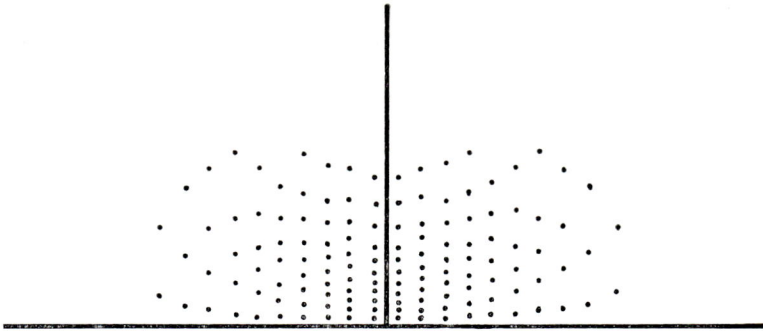

Fig. A.10.1. Pepper-pot diagram for discomfort glare; line-of-sight perpendicular to window.

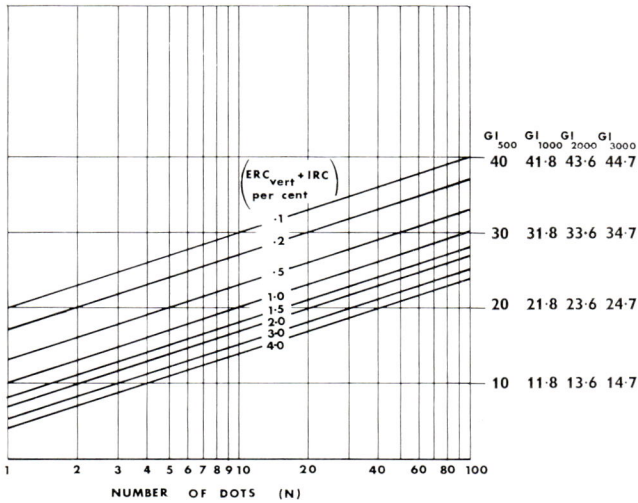

Fig. A.10.2. Glare Index as function of *N* and (ERC$_{vert}$ + IRC); line-of-sight perpendicular to window. To measure ERC$_{vert}$ use Fig. A.9.3 or A.9.7.

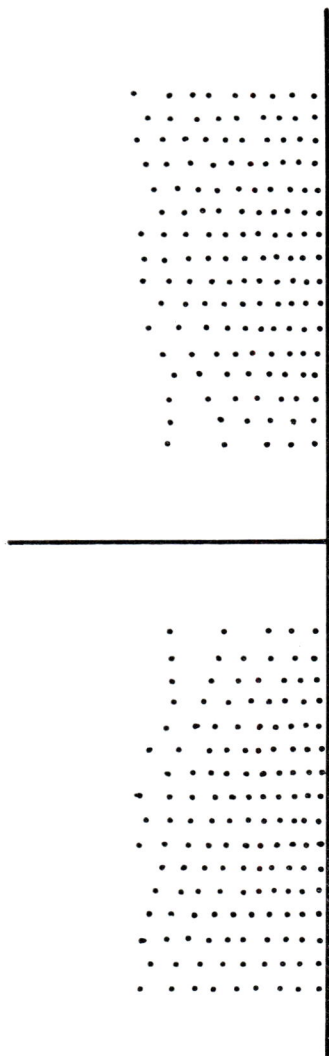

Fig. A.10.3. Pepper-pot diagram for discomfort glare; line-of-sight parallel to window.

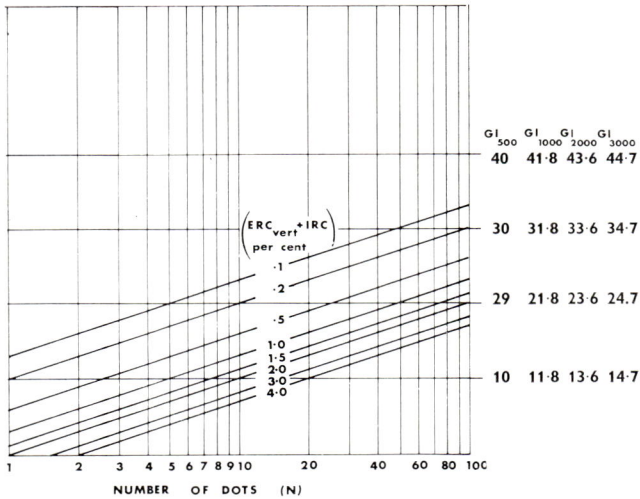

Fig. A.10.4.   Glare Index as function of *N* and (ERC$_{vert}$ + IRC); line-of-sight parallel to window. To measure ERC$_{vert}$ use Fig. A.9.4 or A.9.8.

Index

209